RESTORED
TO
EARTH

RESTORED

TO

EARTH

Christianity, Environmental Ethics,
and Ecological Restoration

GRETEL VAN WIEREN

Georgetown University Press
Washington, DC

Library of Congress Cataloging-in-Publication Data

Van Wieren, Gretel.
    Restored to earth : Christianity, environmental ethics, and ecological restoration / Gretel Van Wieren.
        pages   cm
    Includes bibliographical references (pages) and index.
    ISBN 978-1-58901-997-3 (pbk. : alk. paper)
    1. Human ecology—Religious aspects—Christianity.   2. Environmental ethics.   3. Restoration ecology.   I. Title.
GF80.V38 2013
261.8'8—dc23
                                                                                2012042487

∞ This book is printed on acid-free paper meeting the requirements of the American National Standard for Permanence in Paper for Printed Library Materials.

15   14   13   9   8   7   6   5   4   3   2   First printing

Printed in the United States of America

CONTENTS

PREFACE

Volunteer Revegetation

After all the stabbing at
slick clay soil in the rain,
and all the hands, backs,
eyes, knees, working the plants in
in the mud,
and after passing pots, picks,
spades, cups, bagels, chuckles, shovels,
and so many how-to's and
how-come's and
all too simples explanations,
the last thing we did was back out,
driving live stakes into the ground as we went,
erasing our own
route in.

If it is the path
that makes the garden,
and the garden that
civilizes the wild,
we are disengardening now,
turning on our past
and our pioneering ways
to make amends for the
scythe that went too far,
to say a thank you
audaciously
for the future.

—Cindy Goulder[1]

Recently, over dinner, a colleague and friend of mine expressed frustration about a keynote address that we had both just attended by a prominent ecological theologian. My friend was lamenting what he perceived as an overly negative, crisis-oriented, and dour message that ran through the talk, as well as through contemporary environmental ethics in general. He wondered if mourning the environmental crisis and our bad environmental behavior was really going to motivate better environmental actions. Was a professional theologian telling us why we ought to be more ecologically thrifty, self-sacrificial, and virtuous really going to make us so? Were there not more positive, solution-oriented, and inspiring stories that could be told about how we as modern human beings might live more cooperatively and graciously with land?

This project is, in part, an attempt to respond to these questions. It investigates the human relationship to nature, including ethical and religious values in relation to natural landscapes, from within the context of a particular healing ecological practice. The arguments that follow are based on my perception that much of contemporary environmental ethics, especially religious environmental ethics, has been based on a crisis-oriented approach to environmental problems and ethical responses. While this approach may have had its time and place within environmental ethical thought, and still may have a place in certain circumstances, the time has come for environmental ethics to move toward more positive, forward-leaning approaches and solutions to environmental issues—not toward a romantic optimism that overlooks the gravity of the world's escalating environmental problems, rather toward a course charted between lament and hope.

Among the most distinctive contributions of ecological restoration is its simultaneous focus on degradation and healing, impoverishment and empowerment, loss and recovery. "All is not well in the land community," most restorationists heartily acknowledge; yet they also recognize, firsthand, that "all can be well," or at least, "more well" in terms of health in land and the human relationship to land. Restoration, as I state persistently throughout this book, lives in that creative, paradoxical place between despair over what has been destroyed and gladness that it is still possible to give back to land the capacity for self-healing and renewal. As the poem at the outset states: "*we are disengardening now, turning on our past and our pioneering ways to make amends for the scythe that went too far, to say a thank you audaciously for the future.*"

This book argues that fundamental, significant, and lasting environmental change will occur only when avenues are created for people to physically, intellectually, socially, and spiritually connect with the natural world. Accordingly, it assumes that the generation of environmental virtues—attentiveness, respect, admiration, care, and love—are rooted primarily in our experiences of nature. Human beings can alter their minds, an intellectual act, but they must also change their hearts, an emotional endeavor, for real change to occur. Before we can act ethically in relation to earth we must experience an obligation, and a fundamental desire, to act in a particular way. We must, in other words, feel and know something from within, which is often inherently tied to something experienced from without. The experiences of touching, feeling, and participating in the processes and ways of the natural world will need to form the fundamental basis of a land ethic. It is these basic experiences that the work of ecological restoration can provide.

The ideas put forward here would not have reached book form without the considerable support, information, inspiration, and goodwill of numerous people. I want to thank my wonderful teachers at Yale, Margaret Farley, Tom Ogletree, Gene Outka, Steve Kellert, Baird Callicott, and Holmes Rolston, whose particular perspectives on the study of religion, ethics, and the environment helped to shape, and have their mark on, this project throughout. To Margaret in particular go my deepest admiration, respect, and affection, for lending her clear ethical thinking, profound insight, sharp editorial eye to the manuscript in its entirety, and for her friendship throughout. Many thanks go also to Willis Jenkins, who read and provided feedback on the whole manuscript, and to Bill Jordan and the Values Project Roundtable members for numerous stimulating conversations about ecological restoration practice, ritual, and environmental values.

The initial stages of this project were supported financially by Yale University and the Louisville Institute, and I am grateful for their aid. I completed and revised the manuscript during my first years of teaching at Michigan State University, where colleagues in my home department of religious studies, as well as in the broader university, have welcomed and encouraged cross-disciplinary thinking about environmental ethics. At Georgetown University Press, many thanks to Richard Brown for his early enthusiasm for the project for and his guidance—and that of his entire staff—through the publishing process.

I owe special thanks to those who, offering their time and goodwill, served as guides through the on-the-ground realities, struggles, and joys of restoration work: conversations with Marty Illick from the Lewis Creek Association, David Brynn from Vermont Family Forests, Sister Mary David Walgenbach from Holy Wisdom Monastery, and Gene Bakko from St. Olaf College provided invaluable insights and substance for the heart of this book. And finally, innumerable thanks to my family, Jeff, Inga, Clara, and Carl, for their willingness to help put in a little prairie of our own and mostly for their unending love and encouragement through it all.

NOTE

1. Goulder, "Volunteer Revegetation." Originally published in *Ecological Restoration* 14.1 (1996): 62. © 1996 by the Board of Regents of the University of Wisconsin System. Reproduced by the permission of the University of Wisconsin Press.

❧

FROM WOUNDED LAND AND SPIRIT TO HEALING LAND AND SPIRIT

The Significance of Ecological Restoration for Environmental Ethics

We care for the land because it is good for the land.
We care for the land because it is good for the Lake Mendota
watershed.
We care for the land because it is good for the souls of all God's
people.

—Holy Wisdom Monastery, Middleton, Wisconsin

Environmentalists have long linked the modern environmental crisis with a crisis of the human spirit—of consciousness, the personal heart, or soul. Some thirty years ago essayist and poet Wendell Berry called the ecological crisis a crisis of character.[1] More recently, ecological theologian Mark I. Wallace writes that the global environmental crisis "is a matter of the heart, not the head ... we no longer experience our co-belonging with nature in such a way that we are willing to alter our lifestyles in order to build a more sustainable future."[2] What we need to do, proposes social ecologist Stephen Kellert, is "address the roots of our predicament—an adversarial relation to the natural world—and find a way to shift our core values and worldviews not just toward the task of sustainability, but toward a society with a meaningful and fulfilling relationship with the creation."[3]

The question, of course, is what on earth is going to bring about the transformation that is needed; what is going to help us, once again and anew, find our place and purpose within this beautiful, generative earth? One response that has been frequently overlooked by scholars of environmental ethics, especially religious-oriented environmental ethics, is that of ecological restoration. Ecological restoration is the attempt to heal and make nature whole through the science and art of repairing ecosystems that have been damaged by human activities. It "involves all manner of work with the land, from removing roads and restoring the contours of terrain to removal of exotic plant and animal species that are eliminating or outcompeting native species; to planting trees, grasses, and wildflowers; to propagating endangered plant and animal species."[4] Restoration projects range from the massive, multibillion-dollar Kissimmee River project to restore more than twenty-five thousand acres of Everglades' wetlands to the $30 million effort to restore selected wetlands in industrial Brownfield sites of Chicago's south side Lake Calumet to the reintroduction of tall grass prairie ecosystems in various communities in the Midwest to reforestation and tree-planting efforts throughout eastern Africa.

But beyond this, ecological restoration is the attempt to heal and make the human relationship to nature whole. In its metaphysical understanding of the fundamental interconnectedness of nature and culture and in its practice that provides an experiential bridge between people and land, ecological restoration is viewed by its proponents as providing a promising and moral model for human living with land. In the actual practice of repairing degraded lands—reintroducing, reforesting, revegetating, ripping out, and so on—people and communities are, in an important sense, restored to land.

Additionally, ecological restoration as a healing practice is understood as a form of restitution for past (and present) unjustified destruction and exploitation of land and land-based communities. It also serves as an important vehicle of empowerment for communities whose native ecosystems have been degraded or damaged for the purpose of cultural and economic progress. In this way ecological restoration "is part of a larger mission to create a society that respects democracy, decency, adherence to the rule of law, human rights, and the rights of women," as Wangari Maathai, Nobel Peace Prize winner and founder of the Green Belt Movement, wrote.[5]

One of the unique aspects of the ecological restoration movement in comparison with other conservation or preservation efforts is its considerable utilization of volunteers. For example, the Chicago Wilderness project, which has thus far restored more than seventeen thousand acres (with an additional eighty-three thousand acres in the long-term management plan) of the region's historic oak savanna landscape throughout the city of Chicago, utilized upward of three thousand volunteers annually at its height. The Midewin National Tallgrass Prairie project in Joliet, Illinois, the former site of the Joliet Army Ammunition Plant, which required extensive cleanup from contamination from decades of TNT manufacturing and packaging, relied heavily on volunteers to plant its more than fifteen thousand acres of tall grass prairie. Further, as evidenced in the Kissimmee River and Lake Calumet projects mentioned above, large amounts of public funds are being directed toward restoration efforts that are increasingly understood as vital to promoting a region's ecological health. These two indicators of the importance of environmental activities—number of volunteer hours logged and amount of funds spent—are making ecological restoration one of the most critical and significant environmental priorities on the national, and even the global, environmental agenda.[6]

Even more than a major environmental priority, ecological restoration is understood by some authors as a new paradigm for envisioning the human relationship to nature. William Jordan, founder of the first ecological restoration practitioner's journal (originally *Restoration and Management Notes*, now *Ecological Restoration*) and longtime staff member at the famed arboretum at the University of Wisconsin at Madison, argues that ecological restoration represents the "new communion with nature." Pioneering bioregional activist and writer Stephanie Mills believes that protection of existing wild lands coupled with ecological restoration represents "the main hope that the organic quality of wildness may someday be resurrected in human souls and in all life-places on planet Earth."[7] And religious ethicist Anna Peterson cites ecological restoration as a paradigmatic example of the type of transformative ecological social action that is necessary for the development of a new environmental consciousness.[8] It may be, authors such as these argue, that restoration provides one of the best examples we currently have for rejuvenating ecological health on a planetary scale, for connecting people experientially to the natural world, and for building a culture of nature in which people take care of their natural lands.

The purpose of this book is to examine the significance of ecological restoration thought and practice for environmental ethics, and for our understanding of the human relationship to nature. Inclusion of ecological restoration in environmental ethics is crucial for a number of reasons, and I say more about this below. For now, however, it is important that I say something further about the science of restoration ecology (note the flip in terms). Although this book relies mostly on an understanding of ecological restoration as a movement—that is, the total set of ideas and practices, scientific, social, political, ethical, aesthetic, and spiritual that is operative in restoration projects—scientific ecological concepts and discourse are central both to restoration practice and to environmental ethics, and their examination warrants our attention.

THE SCIENCE OF RESTORATION ECOLOGY

As the definition at this chapter's outset indicates, ecological restoration is both a scientific and an artistic endeavor. Although it began as an enterprise of landscape architects in the nineteenth century, the modern-day ecological restoration movement arose out of the scientific experimental work of Aldo Leopold at the University of Wisconsin at Madison arboretum in the 1930s and 1940s. The arboretum, initiated by Leopold and several of his colleagues, is considered to be one of the oldest and most successful ongoing restoration projects in the country. It was unique at the time for restoring native Wisconsin species and historical ecosystems rather than planting an international variety of species, as many arboretums at the time had done. Furthermore, it was through work at the arboretum that Leopold and his team began to develop what is now known as the scientific field of restoration ecology.

The science of restoration ecology includes various fields such as conservation biology, geography and landscape ecology, wetland management, rehabilitation of resource-extracted lands, and adaptive ecosystem management. Restoration ecologist Eric Higgs proposes the following instructive taxonomy for relating the science of restoration ecology and the practice of ecological restoration, as well as for distinguishing ecological restoration from other related environmental practices such as mine reclamation and revegetation.[9] At the *family level* lies ecosystem management. Under that, at the *genus level*, lie conservation biology, ecological restoration, reclamation, mitigation, and other genera. Finally, at the *species level* fall experimental restoration ecology, rehabilitation, community-based projects, agro-ecosystem restoration, professional projects, and other species of environmental practices.

Using concepts and predictive and mathematical models to explain pattern and processes in ecological systems, restoration ecology develops theories to guide the applied practice of ecological restoration, as well as to advance the academic study of ecology. Although restoration ecology, as with ecological theory generally, does not present a monolithic or unified theory about the structure and function of ecosystems, it nevertheless represents a body of scientific knowledge that addresses virtually every aspect of ecological interaction—from issues related to gene flow to population dynamics to community composition to food-web networks to diversity-stability relationships to large-scale ecological processes, and species and population migrations over time and space.[10]

Restoration ecology is often referred to as the acid test of ecological theory. "If we really understand how ecosystems are constructed and how they function," in other words, "we should be able to put them back together and make them work again."[11] Further, restoration sites are often understood as virtual playgrounds for ecologists, providing an ideal laboratory setting for "asking how well ecological theories can predict the response of natural systems."[12] Areas of ecology that are foundational to the science of restoration ecology include population and ecological genetics, community ecology, evolutionary ecology, food webs, biodiversity and ecosystem functioning, invasive species and community invisibility, macroecology, climate change, and research design and statistical analysis.[13] These focus areas help to generate key research questions for experimental restoration inquiry, including: How can we tell if populations will persist? What assemblages will persist in each part of the restored site? In what order should they be introduced? Can a single restoration site maximize species richness and ecosystem functions? Can restoration be planned within the context of expected global change?[14]

The importance placed on scientific ecological inquiry in restoration practice raises additional issues related to epistemology (the study of the nature of knowledge and justification).[15] In particular it highlights questions regarding the nature of scientific claims about ecosystems: Should scientific claims be considered objective and other types of claims, such as aesthetic and ethical ones, subjective? And how are scientific claims about natural lands to be understood in relation to other types of claims in restoration practice?

Religious environmental ethicist Daniel Spencer addresses the problem of epistemology and the natural sciences in his essay "Restoring Earth, Restored to Earth."[16] Following the work of feminist historians of science Donna Haraway and Sandra Harding, Spencer argues that scientific claims about ecosystems are neither wholly value free nor wholly socially constructed. "Science is always a human practice and discourse that necessarily reflects the social and ecological locations that produce it," he writes.[17] Yet it is not finally reducible to these. Science, Spencer suggests, is importantly "about producing credible accounts of the natural world" that are grounded in concrete ways of life.[18] This does not mean that scientific claims about the natural world should be adopted uncritically, according to Spencer. For so-called objective narratives of science and nature "too often hide the values, assumptions, and power relationships that produced them and may inadvertently be reproduced by them."[19] What is needed in the context of ecological restoration, Spencer proposes, is a hermeneutics of suspicion, a critical approach to science that seriously considers empirical *and* social claims about the natural world.

Spencer's account of science is resonant with the strategic essentialist view of nature that I develop in chapter 2. Yet it also overlooks additional sets of issues and questions that require further analysis in the development of a restoration-based environmental ethic. For example, the issue of how various epistemological claims and narratives should be considered alongside one another in community-based practices such as restoration will need to be considered.[20] This raises further related questions, such as: If ecological restoration practice is fundamentally a scientific endeavor that utilizes empirical knowledge to generate ecological values, to what extent and in what ways should other types of inquiry, knowledge, and values be considered? Moreover, where different epistemologies of nature compete, how should claims be evaluated and weighed?

We will continue to revisit questions such as these in the pages that follow. For now it is important to note that, although the ongoing development of the science of restoration ecology is imperative for the creation of ecologically sound

restoration projects to remedy global ecosystem depletion and its associated problems, the ecological restoration movement should not become exclusively scientific and technical in its approach. Restoration, I argue throughout this book, may
be viewed as a positive and constructive conservation practice precisely because
it has a distinctive capacity to generate multidimensional environmental values,
including communal, aesthetic, spiritual, and moral ones in relation to land.

The extent to which a restoration project uses scientific and technical knowledge versus other types of knowledge depends on the particular goals and objectives of the project, as well as on the type of ecosystem under consideration. A
community-based prairie restoration project in rural Illinois, for example, may
emphasize the value of practical, hands-on experiential types of ecological understanding, whereas a large-scale wetlands restoration project in urban New Jersey
may stress the importance of scientific and technical types of ecological expertise. Nevertheless, for ecological restoration practice to contribute to a more fully
flourishing societal restoration ethic, it will need to be solidly grounded in ecological science and practical experience. We turn now to some definitional clari-
fications. For if part of this book's purpose is to develop a restoration-based ethic,
advancing religious environmental ethics in a more action-oriented direction, it is
necessary that we define more clearly what is meant by the term "environmental
ethics," including how broadly or narrowly it is to be understood and what key
topics fall under its rubric.

DEFINING ENVIRONMENTAL ETHICS

The category of environmental ethics refers to a far-ranging set of issues and
questions related to the moral aspects of the human relationship to the natural
world. Historically, these questions have emphasized the need to define the concept "nature" in relation to human nature, culture, and the sacred. Classical philosophies of nature such those developed by John Stuart Mill and Ralph Waldo
Emerson addressed the question of whether humans could, and should, follow
nature (*naturam sequi*) in any way and, if so, how.[21] Similarly, historical theologies of nature such as those articulated by Thomas Aquinas addressed the issue of
natural law and, specifically, whether nonhuman beings, along with human beings,
were imprinted with the divine law and therefore participants in God's eternal
providence.[22]

In contemporary terms the category "environmental ethics" refers to a field
of study developed in the early 1970s by professional philosophers and religion
scholars in response to the environmental crisis. The publication of Rachel Carson's
Silent Spring (1962) as well as the Sierra Club campaign against the proposed dam-
ming of the Colorado River in the Grand Canyon National Park (1965–68) had
generated widespread public concern about ecological degradation. Additionally,

historian Lynn White Jr.'s famous essay in the journal *Science*—"The Historic Roots of the Ecological Crisis" (1967)—prompted questions regarding the crisis's underlying, fundamental roots. These writings and events, in turn, prompted philosophers and theologians to again, though in the context of a new set of societal problems, begin asking fundamental questions about the moral dimensions of the human relationship to nature.

Although theorists initially focused on questions of nature's value and the need to develop a new, nonanthropocentric ethic, topics discussed within environmental ethics today range from conceptual issues regarding nature's meanings and the relationship between environmental science and practical experience to practical issues addressing biodiversity, habitat loss, climate change, wilderness, deforestation, ecosystem management, toxics and pollution, social justice, sustainability, new urbanism, agrarianism, and ecological restoration. Within the nexus of topics considered under the heading "environmental ethics," some authors believe that it is most important to focus on theoretical issues regarding the moral status of nonhuman nature, whereas others emphasize the need for analysis of the political and economic system and the underlying logic of domination that have given rise to the environmental crisis in the first place. Still other environmental ethicists have focused on the human experience of nature and the ways in which these experiences may form ecological values and virtues in relation to natural lands and civic communities.

Religion and Ecology/Nature and Environmental Ethics

Many authors have understood the human relationship to nature and environmental ethics in religious terms.[23] According to these perspectives the environmental crisis is viewed not only as a moral crisis but also, fundamentally, as a spiritual crisis. Although the roots of the crisis may (or may not) lie in a religious (Christian) worldview as Lynn White Jr. argued, its solution for some environmental ethicists, nevertheless, requires a religious response.[24] An environmental ethic requires psychological–spiritual or sacred grounding, according to this understanding, for deep and lasting change to occur. Scientific understandings may be able to tell us what to do with respect to ethical environmental concerns, but religious understandings, these authors believe, can tell us why we should take this or that action.

It is important to note that religion in general and Christianity in particular have not always proven to be a helpful source for promoting earth care. In fact, quite the opposite: Western Christianity's history of colonization and domination of peoples and nature in God's name makes one wonder whether it, or any other religion with a colonizing destructive past for that matter, is redeemable in terms of its ability now to function as a genuine earth faith. Moreover, many authors, including environmentalists such as Henry David Thoreau, John Muir,

Aldo Leopold, and Rachel Carson, as well as theologians such as Joseph Sittler, Rosemary Radford Ruether, Leonardo Boff, Sallie McFague, Larry Rasmussen, and Ivone Gebara, have critiqued Christianity's anti-nature bias and associated environmental attitudes and behaviors.[25]

Central to many of these critiques is the idea that Christianity, especially since the sixteenth-century European scientific revolution and Protestant Reformation, has perpetuated an anthropocentric and a static view of creation and an individualistic and otherworldly understanding of salvation. This worldview has contributed to unethical environmental attitudes and behaviors, these authors argue, representing an overall lack of explicit concern and respect for the well-being of the whole community of life on earth. As theologian Joseph Sittler wrote several years prior to Lynn White's critique, "The doctrinal cleavage, particularly fateful in western Christendom, has been an element in the inability of the church to relate the powers of grace to the vitalities and processes of nature."[26]

Many ecological theologians, including Sittler, also contend, however, that there are resources embedded deep within Christianity (as well as within other religious traditions) that can be retrieved and reformed in order to defend earth and to help form holistic visions for living more justly and peaceably with one another and with the earth. This is not to suggest that obstacles and uncertainties do not remain with regard to Christianity's (or other global religions') naturalizing potential, these writers aver. Rather, that religious beliefs and practices, as part of evolving human cultures, can change; and history has shown this to be true. Where religious beliefs and practices have not adequately reflected or resonated with people's lived experiences or understandings of the divine or interpretations of scripture or whatever, (some) individuals and communities have agitated and advocated for changes within their traditions. Of course, such advocacy is not always successful in terms of bringing about desired changes. Human power, even, perhaps especially, within religious institutions, does not easily give itself over to dissent. Nonetheless, the point here is that religious beliefs and practices are and have always been subject to ongoing change, and there is no reason to believe that this will differ with regard to religious understandings about the natural world and humanity's place and role within it.

Scholars who include a religious perspective of environmental issues emphasize a variety of methods and topics. Some scholars, for example, draw primarily on the sources of specific religious traditions (e.g., sacred texts, theological doctrines) in order to critique, retrieve, or reconstruct these traditions' moral ecological worldviews. Representative of this approach is the work of Mary Evelyn Tucker and John Grim, conveners of the Forum on Religion and Ecology (FORE) and editors of the Harvard Center for the Study of World Religions' ten-volume book series, *Religions of the World and Ecology*.[27] Building on the thought of cultural historian and ecological theologian Thomas Berry, Tucker and Grim's work emphasizes the need to expand "the growing dialogue regarding the role of the world's religions as moral forces in stemming the environmental crisis."[28] In doing so they also encourage the global religions to tap their traditions for sources of wisdom that may help spiritually narrate the new scientific universe story of cosmic and earth evolution.[29] The global religions "remain one of the principle resources for symbolic ideas, spiritual inspiration, and ethical principles," write Tucker and Grim. It may be that they require critical evaluation, reform, or even reconstruction with regard to their views of earth and humanity's role within it, yet the point remains: religion plays an integral role in forming cultural values and worldviews, which in turn shape a society's ethical orientation toward the natural world.[30]

Tucker and Grim recognize that there are limitations with regard to religion's capacity for effecting environmental change. Institutional religion has a dark side, as we have noted, and this causes pause when it comes to inciting its socially transformative potential. Religions have been "the source of enormous manipulation of power in fostering wars, in ignoring racial and social injustice, and in promoting unequal gender relations, to name only a few abuses," they write.[31] Furthermore, even in cultures where religious views have appeared to be sympathetic to nature, there has often existed significant environmental destruction.

This "disjunction between the ideal and the real," however, as well as religion's shadow side, "should not lessen our endeavor to identify resources from within the world's religions for a more ecologically sound cosmology and environmentally supportive ethics," state Tucker and Grim.[32] Rather, it should give rise to a process of reflection whereby the religions' conceptual resources are explored and evaluated and people's perspectives are challenged and broadened.[33] It may be that the complexity of environmental problems "requires interlocking approaching from such fields as science, economics, politics, health, and public policy," they write, though it is nevertheless "inevitable that we will draw from the symbolic and conceptual resources of the religious traditions of the world" in our search for "more comprehensive ecological worldviews and more effective environmental ethics."[34]

Other religion scholars look to sources outside institutional religion, citing the ways in which nature-based activist or recreational groups often link understandings of nature's sacred value with an environmental ethic of respect and care.[35] Most notable is the work of Bron Taylor, founder of the International Society for the Study of Religion, Nature, and Culture (ISSRNC) and editor of the multivolume *Encyclopedia of Religion and Nature* (ERN). The ERN was conceived to remedy "the lacunae in the inherited [from Tucker and Grim] 'religion and ecology' field," states Taylor in its introduction.[36] The lacunae run along two interrelated lines, according to Taylor: marginal or boundary-transgressing nature-based spiritual experiences have been neglected, and methods appropriate for examining them have been ignored. To correct these oversights Taylor and others have focused their "religion studies lens" on the lived-nature religions of those who follow a radical environmental or "dark green" worldview.[37]

Influenced heavily by the lived-religion approach of religion scholars Robert Orsi and David Chidester, this wing of religion and ecology (termed "religion and nature" by Taylor) uses methods more historical/sociological in mode than confessional/ethical (the latter representing Tucker and Grim's approach, according to Taylor). Empirical–scientific (historical/sociological) approaches to the study of religion and ecology are preferred in this approach, presumably because they offer a better, more precise indicator of the way in which religious beliefs and practices influence environmental attitudes and behaviors. To this end adherents of this approach often favor methodologies such as ethnography and social scientific surveys.[38] Groups such as Earth First!, homesteaders, surfers, kayakers, and fly-fishers, for instance, are among those examined by those committed to Taylor's lived-nature religion approach.[39]

I draw on both of these approaches to the study of religion and ecology/nature in the environmental ethic constructed in this book. Although I analyze sources within the Christian tradition (Tucker and Grim's approach), I examine religious elements embedded in the ecological restoration movement and the fundamental experience of the human connection to nature that restoration activities may yield (Taylor's approach).[40] I say more below about what I mean by the concepts of religion, spirituality, ethics, and morality. Suffice it to say here, the environmental ethical approach I use in this book is best thought of as religious environmental ethics, given my attention to the Christian tradition and to the spiritual dimensions of nature-based experience.[41]

Nonanthropocentric and Experience-Based Approaches to Environmental Ethics

Related to the debate among religion and ecology scholars is one of the biggest lines of disagreement in environmental ethics; that is, the issue of whether a unified, nonanthropocentric theory of value should be developed, or whether plural, experience-based approaches to particular environmental problems should be emphasized. In philosophical environmental ethics this disagreement is evident in the ongoing debate between intrinsic value theorists (J. Baird Callicott, Holmes Rolston) and environmental pragmatists (Eric Katz, Andrew Light). The debate over nonanthropocentric versus pragmatic approaches to environmental ethics is reflected similarly in religious environmental ethics.[42]

Anna Peterson, for example, argues that the link between how we *think* about ourselves in relation to nature (nonanthropocentric worldview) and what we *do* or who we *are* in relation to nature (environmental ethical action and virtue) may not be as causal or airtight as environmental ethicists have assumed for the past four decades. Environmental ethics, according to Peterson, should "abandon the idealist assumption of a simple and unidirectional relationship between ideas and practice, in which practice is always derivative or secondary to ideas and which believes that if we get the ideas right, then the practices will follow."[43] It is not that theorists should discontinue the task of clarifying and redefining ideas, writes Peterson; rather, they should reorient the task and method of environmental ethics to focus on lived environmental experience.[44]

Despite what I think are compelling reasons for developing an environmental ethic based on lived experience and plural moral ecological values, I will assume in this book that both nonanthropocentric and experience-based approaches are important and necessary in developing a restoration environmental ethic. For when concrete environmental practices such as ecological restoration are examined, we find that an emphasis on conceptual and methodological divides such as the one just cited may not be as necessary as some scholars have thought them to be. Consequently, upholding these divides may prove counterproductive for reaching greater understanding about the human relationship to the natural world.

Furthermore, there remains a significant level of uncertainty within environmental ethics when it comes to understanding how cultural change actually occurs and, therefore, which ethical theories should be privileged based on their potential for creating positive environmental action. As I argue in later chapters, ethical perspectives that emphasize a nonanthropocentric outlook are not necessarily incompatible, for instance, with those that seriously consider the role of human experience and civic virtue. This becomes especially apparent when we see the ways in which many restorationists draw on both nonanthropocentric and anthropocentric meanings of nature to justify ecological restoration activity and describe the human relationship to land.

A recurring set of questions emerges in the matrix of environmental ethics, despite the multiplicity of issues and topics that fall under its heading. These include, for example: Can a nonanthropocentric or holistic environmental ethic adequately account for the needs of individuals, human and nonhuman? Should considerations of human justice and well-being trump those of ecological integrity and health? Do new evolutionary understandings of nature and humanity require revision or replacement of traditional philosophical and religious resources, or can historical sources simply be critically retrieved? Should an environmental ethic be grounded in scientific or religious—or both—understandings of nature and humanity? Is the development of a global environmental ethic possible and desirable, or is the generation of environmental ethics necessarily a place-based activity?

Many of these questions are addressed throughout this book. Next, however, another preliminary step is in order. Despite the fact that ecological restoration has played an insignificant role in contemporary environmental ethics, as already noted, restoration issues have nevertheless been attended to by a small group of scholars in the field. I turn next to examine the ways in which ecological restoration *has* been explicitly considered in environmental ethics, noting specifically the literature that will be useful for the restoration ethic developed in this book.

ENVIRONMENTAL ETHICS AND
ECOLOGICAL RESTORATION

Treatments of ecological restoration in contemporary environmental ethics have, over the past several decades, been few and far between. Interestingly, however, where restoration discussions have developed, they have touched on a dizzying array of critical environmental topics, from the value of nature to the relationship between culture and nature, to the role of civic experience in environmental stewardship, to the moral status of wild versus domestic, indigenous versus nonindigenous organisms. Contributing to this plurality of topics, environmental ethicists have offered mixed sentiments with regard to restoration's legitimacy as a conservation practice. Although specific considerations of ecological restoration within environmental ethics have been limited in number, they have, nevertheless, been expansive in scope. Thus, to focus key issues and questions with regard to ecological restoration and environmental ethics, I begin by examining relevant work from philosophical environmental ethics, moving next to religious environmental ethics, and finally to ecological restoration.

Philosophical Environmental Ethics

Among the first environmental ethicists to seriously consider the topic of ecological restoration were philosophers Robert Elliot and Eric Katz. Elliot and Katz, writing in the 1980s and '90s, did not formulate a positive assessment of restoration as an environmental ethical practice; rather, both argued in vehement opposition to what they perceived as restoration's underlying anthropocentric ontology. As we shall see in detail in chapter 2, ecological restorations, according to Elliot, were "faking nature," whereas according to Katz, they were a "big lie."[45] Culture is culture and nature is nature, argued Elliot and Katz—and they should remain this way: separate and distinct, especially for the sake of protecting nature's inherently wild ways. Human activities may be (more or less) aligned with evolutionary processes, though restoration, according to these early theorists, was clearly not an example of this. Restoration was an unwelcome human meddling in natural systems.

Shortly after Elliot and Katz developed their anti-restoration arguments, several environmental philosophers attempted to redeem restoration's value as a conservation practice. Most vocal among those who came to restoration's rescue in the mid-1990s was environmental pragmatist Andrew Light, who believed that professional philosophers needed to present a more positive view of restoration, lest the restoration community come to believe that philosophers had only a negative take on a social practice that was at the time gaining prominence within the broader conservation movement.[46] Light therefore pursued restoration's potential as a civic, community-building activity rather than the ontological issues initiated by Elliot and Katz. Furthermore, Light argued that environmental philosophy as a discipline was in need of a pragmatic restoration—salvation, if you will, from what he and others viewed as an overemphasis on intramural meta-debates regarding the real value of nature—which ecological restoration could help promote.

Yet it was not only pragmatic environmental philosophers such as Light who attempted to redeem restoration's value as a conservation practice. Interestingly, intrinsic-value theorist Holmes Rolston also came to restoration's rescue. Although Light opted to focus on more typically pragmatic arguments related to civic experience, Rolston used ones about nature's intrinsic value to counter those put forward by Elliot and Katz. According to Rolston, humans can and should restore ecosystems that have been damaged by human activities as a way to give back natural values to the land community. Restoration can be viewed as an altruistic act, Rolston proposed, one of moral restitution "for the sake of the wild others who may re-reside there."[47] Although humans should restore for nature's sake, ecological restoration, according to Rolston, can also nevertheless "increase our human sense of identity with nature" and help us "appreciate the biotic community we have studied and helped to restore."[48]

But how is this complex mix of ecological and human, intrinsic and anthropocentric values that Rolston describes to be articulated in an environmental ethic? Moreover, both Light's and Rolston's proposals make one wonder whether there might be additional, perhaps equally important, restoration-based experiences, values, and virtues that environmental ethicists have overlooked. For example, might restoration activities generate certain spiritual experiences and moral virtues in relation to damaged and healing natural lands? Are there particular human experiences, most notably those of marginalized groups of people, that have been overlooked in dominant restoration thought? And have pragmatic treatments of restoration such as Light's neglected to consider adequately the ways in which some restorationists continue to be motivated by an ethical, even religious notion that restoration activities are an expression of respect for nature's intrinsic, sacred value?

Religious Environmental Ethics

Religious environmental ethicists have come only recently (and far more infrequently than philosophers) to the restoration discussion, despite the fact that many restoration questions such as the ones just noted invite further religious and ethical analysis. Only two essays have been written specifically on the practice of ecological restoration in the field of religious environmental ethics. Both have been written in relation to the Christian tradition in particular, despite their differing perspectives on restoration's potential as a moral ecological model.

Christian ethicist Michael Northcott critiques dominant restoration thought and practice for its neglect of justice-related concerns.[49] Northcott's argument, which I evaluate in chapter 4, focuses on a particular wilderness restoration project being undertaken in his home region near the Cairngorm plateau wilderness area of the Scottish Highlands. The project, believes Northcott, has wrongly attempted to "repristinate" the ecosystem as if humans, indigenous or European, were never inhabitants of the landscape. This type of restoration effort, along with dominant restoration philosophical thought more generally, fetishizes wild nature in its view that nature needs to be protected from human dwelling. Further, and more significantly, restoration philosophers such as Elliot, Katz, and Light have failed to adequately consider justice-related concerns, according to Northcott. They provide "no answer to the situation of many Highland communities," Northcott writes, "where indigenous children can no longer afford to settle because of the high market value of homes with views of, or proximate to, wilderness areas such as the Cairngorm Plateau."[50]

Unlike Northcott, Daniel Spencer argues that ecological restoration can provide a beneficial and moral model of human living with natural lands. Spencer draws primarily on the work of restorationists Eric Higgs and William Jordan, as well as religious environmental ethicist Larry Rasmussen, to develop a restoration-based ethic for reinhabiting place.[51] Following the inclination of bioregional restorationists, which I say more about below, Spencer suggests that the activities of restoration may help humans reinhabit their natural landscapes in socially and ecologically positive ways. These activities may also serve as a framework, writes Spencer, for building an ethic in which human environmental values are generated and refined through particular healing actions. A restoration ethic "is an ethics of praxis," Spencer proposes. "It begins by healing the humanity-nature rift by immersing ourselves in creatively reinhabiting our places through acts of restoration."[52]

Both Northcott's and Spencer's arguments resonate with the ones advanced in this volume. Yet each is in need of further analysis, clarification, and development. Northcott's concepts of restoration and of justice, as I argue in later chapters, are limited in scope and require consideration of additional concepts and issues. Moreover, Spencer's proposal points to important methodological and epistemological issues (noted above with regard to the science of restoration ecology) that necessitate deeper examination if ecological restoration is to serve as a potential framework for an environmental ethic. The notion, for instance, that religious and ethical values are generated and formed in and through transformative human actions has never been accepted readily in dominant philosophy or theology. And the view that spiritual and moral values may be generated in and through working with natural processes and systems raises the perennially debated question in Christian thought regarding the relationship between nature and grace, and human freedom and divine action. These are among the complex set of religious and ethical issues that will need to be addressed in subsequent chapters. Yet additional treatments of the religious and ethical elements of ecological restoration must be noted as well.

Ecological Restoration, Spirituality, and Ethics

Beyond professional ethicists such as the ones just cited, several key authors in the field of ecological restoration have explored the practice's religious and ethical implications. Bioregional pioneers Freeman House and Stephanie Mills have popularized restoration spiritual and moral potential through their beautiful, poetic treatises, *Totem Salmon: Life Lessons from Another Species* and *In Service of the Wild: Restoring and Reinhabiting Damaged Land*, respectively.[53] House and Mills, as I expand on in various places in the book, see restoration as a form of reinhabitation, a way to enter into ongoing and long-term relationships with particular landed places (bioregions). Restoration as reinhabitation implies living in, and having an economic stake in, the place restored. "It means the articulation of a culture of place, and for this to come about in our time means the restoration of human community in a society whose members have wildly differing and fiercely held ideas about what land is for."[54]

In the academic ecological restoration literature, restoration's spiritual and moral dimensions have been discussed in important ways by authors such as Dennis Martinez and Eric Higgs. The work of Martinez—cofounder of the Indigenous Peoples' Restoration Network (IPRN), a working group of the Society for Ecological Restoration (SER)—has been significant in terms of its focus on the link between ecological restoration and traditional ecological practices (TEK) in indigenous cultures. I repeatedly reference his work with the IPRN, as well as with the Sinkyone Intertribal Park restoration project in northwestern California, given its emphasis on the intersection of indigenous spirituality and ecological restoration.[55]

Restoration ecologist and past president of the SER Eric Higgs includes several considerations of restoration's spiritual and moral implications in his volume *Nature by Design: People, Natural Processes, and Ecological Restoration*.[56] Specifically, Higgs proposes a notion of "focal restoration," which I critically analyze in chapter 4, whereby people can form a sense of meaning, connection, satisfaction, and even atonement through the actions of digging into the earth, seeking "a new way of relating to things other than human," or enjoying "the pleasure of good company and good work."[57] Higgs argues more broadly that restoration can provide "a redemptive opportunity," an idea I address in chapter 6. We may, according to Higgs, "heal ourselves culturally, and perhaps spirituality, by healing nature."[58]

Perhaps more than any other figure, however, William Jordan has promoted restoration as a context for creating transformative emotional and symbolic experiences in relation to the land. Considered the movement's leading visionary, Jordan was for twenty-five years a staff member at the arboretum at the University of Wisconsin where he coined the terms "synthetic ecology" and "restoration ecology." He is also the author of the already classic volume *The Sunflower Forest: Ecological Restoration and the New Communion with Nature*.[59] Drawing on the value theory of literary critic Frederick Turner, Jordan has criticized contemporary environmental ethics and environmental thought in general for skipping over the hard edge of value generation in relation to nonhuman nature. In ignoring the inevitably negative (or difficult or shadow) side of the human relationship with the natural world (e.g., removing, culling, killing), environmental ethics, according to Jordan, has fostered an overly romantic view of community. Such a romantic view has precluded the formation of a deeper, truer, more durable type of communion between people and land—a communion that Jordan believes ecological restoration is distinctly poised to promote.

Throughout this book I critically probe each of the above ecological restoration perspectives on spirituality and ethics. Yet, as will become evident as my argument develops, I also go beyond these treatments in significant ways, especially in terms of the range of experiences, emotions, and values I explore with regard to restoration thought and practice. Much more needs to be said about the question of the meaning of ecological restoration, a task I undertake in chapter 1. Next, however, an additional preliminary step is needed. As I have already noted, and also specify in subsequent chapters, additional sets of issues are raised when religious and ethical questions are explicitly considered in a meaning for ecological restoration.

RELIGION, ETHICS, AND ECOLOGICAL RESTORATION

Part of the task of this book is to highlight the neglect of religious elements in understandings of ecological restoration. For even though the spiritual dimensions of restoration have been addressed in the ecological restoration literature, restoration's religious aspects and implications have not been considered systematically and explicitly in the dominant work in the field, for example, in the Society for Ecological Restoration definitions or in environmental philosophical perspectives.

Thus, in this section I raise four potential sets of issues for the meaning of ecological restoration that emerge precisely when religious and ethical aspects are seriously considered. Before I do this, however, it is important that I define the categories "religious" and "ethical," as well as the correlative terms "spiritual and "moral," given their centrality throughout this project.[60]

Defining Religion, Spirituality, Ethics, and Morality

In contemporary terms, spirituality is generally contrasted with religion: spirituality is viewed as personal and subjective and religion is understood as institutional and dogmatic. Sociologists such as Wade Clark Roof tend to accept this distinction, as in: "to be religious conveys an institutional connotation [while] to be spiritual ... is more personal and empowering and has to do with the deepest motivations in life."[61]

Religion and spirituality need not, however, be viewed only in opposition. Both may be viewed as arising from the inherent human impulse to make sense of and find meaning in relation to life's big questions, including the mysteries, wonders, and beauties of the universe and of earth. Although spirituality may be "less oriented to a fixed creed or defined denomination, [and] more committed to the long path toward spiritual truth," this very path, for some, is to be found within the deepest truths of the world's major religious traditions.[62] For example, religious writings for thousands of years have addressed the ways in which religious practices have fostered deep experiences of transcending the conventional self and social norms, communing with the divine or Ultimate Reality in the universe, and feeling a sense of awe, amazement, and wonder in relation to world around oneself.[63]

Within the religion and ecology field, the concept of religion, rather than spirituality, is generally used to refer to nature-oriented spiritual–religious phenomena. Mary Evelyn Tucker and John Grim and Bron Taylor, for example, similarly use the term "religion" in their primary definitions of earth-based religiosity—"religious ecology" and "nature-as-sacred-religion," respectively— despite the fact, as we have seen, that Tucker and Grim focus on analysis of the major global religious traditions and Taylor emphasizes nontraditional and emergent religious activities.

Further, both Tucker and Grim's and Taylor's concepts of nature religion include similar elements. For example, each includes the idea that a religious understanding of the human–nature relationship includes an element of deep connection or kinship between human and nonhuman beings. A religious ecology, according to Tucker and Grim, is characterized by an "awareness of kinship with and dependence on nature for the continuity of all life"; "nature-as-sacred-religion," believes Taylor, involves the idea that humans are "bound to and dependent upon earth's living systems."[64] Moreover, concepts of nature religion such as the ones just cited tend to involve the idea that the human connection with the natural world can foster positive transformation and healing, as well as a deep sense of meaning, satisfaction, and fulfillment. Insofar as nature is understood as in some sense sacred or holy, it is also viewed as "worthy of reverent care."[65] Conversely, in the understandings of nature religion just mentioned, "damaging nature is considered to be an unethical and desecrating act."[66]

As with the notions "religion" and "spirituality," the concepts "ethical" and "moral" are commonly distinguished. "Morality," like "spirituality," tends to be associated with concrete experience and action, whereas "ethical," like "religion," tends to refer to a systematic approach to morality (in religion's case, a systematic and often institutional approach to spirituality). Morality tends therefore to refer to matters of individual human choice and character, while ethics tends to refer to a set of principles or a systematic framework for stipulating what constitutes right or good action, modes of being, or outcomes.

This said, there is no consistent differentiation between moral and ethical in the ethical literature, and I, too, tend to use the terms interchangeably. Still, if a distinction is made between moral and ethical, I tend to follow the difference just noted—that is, I use the terms "ethics" and "ethical" to refer to a systematic approach to morality (environmental ethics) and "moral" and "morality" to refer to the experience involved with deciding to act in this or that way or to be this or that type of person (ecological morality). Similarly, if a distinction is made between religion and spirituality, I tend to use "religion" and "religious" to refer to a systematic approach to spirituality (Benedictines as religious restorationists), and "spirituality" and "spiritual" to refer to concrete experiences or action (restoration-based spiritual experience).

Given that I use a religious and an ethical perspective in this book, I will often join the categories of religious and ethical and spiritual and moral (as in religious–ethical and spiritual–moral) in interpreting ecological restoration's meaning. Moreover, as just noted in regard to nature-oriented religion, I assume that religious belief and experience and ethical action are often intrinsically related. Still, a main contention of this project is that the explicit connections between religion and ethics, spirituality and morality with regard to nature require further systematic analysis—a task that is particularly well fitted with the practice of ecological restoration, for where restoration activities yield religious experiences and understandings of the human–nature relationship, they do so explicitly in and through the ethical action of healing damaged land. In other words, the religious and ethical can be viewed as intertwined right from the start in the practice of ecological restoration.

Considering a Religious, Ethical Perspective

With this terminological excursion in mind, I now want to identify four sets of questions that are raised when a religious ethical perspective is considered in relation to the meaning of ecological restoration. These questions, I argue throughout this book, ought not to be left marginal in ecological restoration thought and practice. My aim here is primarily to identify the categories of these questions in preparation for a fuller response to them in succeeding chapters. The sets of questions relate to the following: (1) the value of nature; (2) transformation of human agents and symbolic action; (3) justice; and (4) the relationship between scientific/secular and religious/sacred understandings. I discuss them singularly and together.

First, consideration of religious ethical dimensions of restoration raises questions regarding the meaning and value of nature. Defining what is, and what counts ethically, as nature has been one of the main preoccupations of contemporary environmental ethicists over the past four decades. Yet the task becomes more complex when we consider the value of nature where land has been damaged and repaired by human practices. What happens to nature's true value when humans degrade ecosystems? Is it lost forever, or can humans, in some sense, help regenerate land's ecological values?

Environmental ethicists have largely overlooked the ways in which notions of nature's standing have been developed within recent ecological restoration thought. This may be because arguments regarding nature's true value were utilized initially by philosophers to oppose rather than defend the practice of ecological restoration. It also may be because environmental pragmatists such as Andrew Light proposed early on that environmental philosophers ought to concentrate their efforts on examining the civic potential of restoration practice rather than on questions of nature's value (lest philosophers devolve into useless meta-babble, according to Light). Additionally, some arguments related to nature's standing within ecological restoration thought border on the religious, a dimension most philosophers and scientists have shied away from in examining environmental issues.

Now that restoration is becoming more widely accepted by environmental ethicists, and popularly practiced by thousands of groups of people globally, it may be worth posing questions related to nature's standing once again (questions that are considered again in chapter 2). Since restorationists have recast abstract theories of nature's value in light of the actual activities of attempting to heal damaged ecosystems, arguments have been reconfigured in novel and illuminating ways; they even utilize religious language, despite restoration's secular scientific basis. For example, some restorationists suggest that the notion of the sacred, historically interpreted in relation to human and divine life, should be extended to nature. This sentiment is implicit in deep ecological–bioregional perspectives on restoration that view nature's spontaneous, self-organizing, and self-renewing processes as intrinsically valuable and, in some sense, sacred, apart from their usefulness to human beings.

Determining which parts or wholes of nature have religious and ethical relevance, however, raises further questions regarding the meaning of nature. For instance, feminist environmental scholars such as Sandra Harding, Carolyn Merchant, Val Plumwood, Sallie McFague, and Rosemary Radford Ruether have raised questions regarding the dominant epistemologies and methodologies used historically (and currently) in developing understandings of nature in the first place. They ask: Why is it the case, culturally and historically, that nature has come to be understood as separate from and less valuable than culture? How has such an oppositional dualistic and hierarchical understanding of culture and nature—as well as male over female, white over black, divine transcendence over divine imminence—contributed to the domination of women and the unjustified destruction of nature? How do scientific assessments of ecosystems reflect this bias? Alternatively, are there different biases held by individuals and communities, secular or religious, that have had longstanding experiences living with land? What assumptions does each of them rely on in coming to such conclusions?

Not all restoration authors are comfortable appealing to intrinsic, let alone sacred, value for nature as a way to justify restoration practice. But, in fact, the debate over nature's essential or constructed value has been central to conversations regarding nature's true value within restoration thought.[67] On the one hand, this reflects a broader trend within academic disciplines over the past two decades to consider critical theoretical, namely, postmodern deconstructionist perspectives on the nature and meaning of reality. On the other hand, restoration practice seems to be especially susceptible to questions related to "what is, and what counts as, nature" given its character as a human activity attempting to work within and through natural processes.

Restorationists are continually faced, for example, with the simultaneous realities that they are in some sense making up nature as they go, based on their own personal experiences and cultural values *and* that natural processes are in some sense other than and beyond the human mind.[68] Efforts to describe, interpret, and emotionally and practically confront this dual reality have challenged restorationists over the past twenty-five years. Some restorationists have stressed the reality of nonhuman natural beings and processes, beyond mental de/construction. Others have criticized what they view as an essentialized, reified understanding of nature or ecosystems, emphasizing instead the contingent and values-based character of these concepts. Still others have advocated for a middle way, which admits the role of human experience and culture in shaping our perceptions of nature, and the role nonhuman natural processes play in the world, apart from the human mind.

As Eric Higgs writes, ecological restoration might best be thought of as a conversation about how humans should live with land. Part of this conversation involves discussions regarding nature's true value, including perspectives that take seriously nature's spiritual–moral meaning. For many indigenous and bioregional restorationists, one of the reasons humans ought to restore damaged land is for the sake of the sacred, wild other(s) in our midst. For many other restorationists, perceptions of damaged and ideal nature are importantly grounded in personal experiences and cultural values in relation to natural systems and beings, including religiously resonating ones in relation to nature's systems and beings.

This leads to a second set of questions raised precisely when religious and ethical dimensions and implications of restoration are considered. That is, what is the religious dimension (if there is one) in the transformation and renewal that many restoration practitioners reference; and what sorts of religiously symbolic actions may restoration perform? As is indicated in the perspectives on ecological restoration described above, restorationists at times utilize religious language to refer to the personal experiences of healing, renewal, communion, and dependency in relation to nature that are generated through the restorative act. Higgs writes that a "special communion forms when people literally dig into the earth to reverse a tide of degradation, atone for past actions, seek a new way of relating to things other than human, or enjoy the pleasure of good company and good work."[69] Deep ecological restorationists such as House and Mills emphasize reinhabiting particular ecosystems as a way to foster deep connections of kinship with all beings, human and nonhuman alike. Martinez states that the Sinkyone people want to restore life, the living and sacred relationship between people and the earth, their spirits, culture, songs, myths, and stories, the Indian names for creeks and springs, their own selves.[70]

But what does all this mean? How, if at all, does the practice of working with a wounded and healing earth shape religious ecological experience? And vice versa, how more concretely can restorationists describe such religious restoration experiences? In what ways do these experiences alter the language of nature-based spiritualities in the world's religious traditions? How do religious restorationists in particular understand the religious experiences wrought through restorative acts?

These questions about the transformative dimensions of restoration experience raise additional ones regarding restoration's symbolic dimensions and its broader cultural and religious meanings (questions to be met again in chapters 5 and 6). Restoration may create spiritual–moral experiences within particular practitioners, but the act of restoration may also help form communal and cultural values in relation to landscapes. Higgs, we recall, writes that "restoration offers a redemptive opportunity. We heal ourselves culturally, and perhaps spiritually, by healing nature."[71] Jordan states that restoration might be thought of as a new ritual tradition, one that has the potential to create a more gracious and durable relationship between people and land. But what do restorationists such as Higgs mean when they say we might be healed culturally through the act of regenerating damaged ecosystems? What language ought we use to describe the restoration of ecosystems that must occur in order for humanity to live in more cooperative relations with land? Does land have a way of forgiving humankind for the damage unfairly wrought to it? If nature is in some sense sacred, what happens to this sacred dimension when ecosystems are damaged, even destroyed? What about when they are repaired?

To respond to questions such as these, some religious ethicists, as already indicated, have argued that scholars ought to focus more on examining particular environmental practices and the values they may shape than on developing further religious or philosophical ecological ideals. Recall Anna Peterson's proposal that environmental ethics should "abandon the idealist assumption of a simple and unidirectional relationship between ideas and practice, in which practice is always derivative or secondary to ideas and which believes that if we get the ideas right, then the practices will follow."[72] Peterson points to the practice of ecological restoration as an example of the sort of transformative, experience- and practice-based activity she views as necessary for generating good or better social environmental values.

Along similar yet divergent lines, religious ethicist Willis Jenkins has called upon environmental ethics to reconsider its task and methods in order to develop more practice-oriented accounts. Jenkins emphasizes the need for religious environmental ethics to go back to the drawing board: it must first determine what counts as an environmental issue and who defines it as such. This is important according to Jenkins because environmental ethics since Lynn White has focused on analysis of the environmental crisis writ large rather than on concrete environmental problems. Religious ethicists would do well, proposes Jenkins, to heed the critiques of the pragmatists in environmental philosophy (e.g., Anthony Weston, Andrew Light, and Bryan Norton) and Bron Taylor in religion and ecology. Combining them, Jenkins proposes that a religious environmental ethic "might admit more plurality and better attend to the adaptive innovations of lived experience (Taylor's admonition) precisely by letting problems determine its normative agenda (the pragmatist counsel)."[73] Yet Peterson's and Jenkins's proposals raise further questions here for religious environmental ethics. For example: What more concretely are the cultural and environmental values that restoration may be able to help shape? How do these values relate to religious and ethical ones, as well as to lived religious experience?

Third, consideration of religious and ethical concerns in relation to restoration raises questions related to the principle of justice. The notion of justice is generally referred to in at least two ways in the environmental ethics literature. These are neither mutually exclusive nor exhaustive, though theorists and activists do tend, in general, to rely more heavily on one over the other. In the first meaning, justice is applied in a particular way to individuals who have been marginalized as groups or in oppressed communities that experience, most acutely, a disproportionate level of environmental "bads" (e.g., pollution, contamination, desertification, deforestation, etc.). This is the predominant understanding of justice of the environmental justice movement.[74] According to this view, the right to a healthy environment is a human and civil rights issue; where environmental injustices exist, they mirror present socioeconomic, racial, ethnic, and gender inequities.

Here, environmental justice refers to the demand for equity in the distribution of environmental burdens and benefits, as well as changes in current social-economic-political structures that create and perpetuate environmental inequities in the first place. Further, justice is understood in terms of recognition: of the particular communities and people that are disproportionately negatively affected by environmental bads, and of the diversity of interpretations and equitable distribution of environmental health and sustainability goods within the environmental movement itself. Justice also requires maximum public participation, according to the environmental justice movement, whereby those citizens and communities who are hurt by this or that environmental activity have a voice in developing and implementing plans, and in forming decision-making procedures and environmental policies that affect their communities.[75] This view of justice is evident in the work of the IPRN, as well as in international development literature in which ecosystem restoration is understood specifically in relation to empowering poor communities to take social environmental political action in order that human (and biotic) livelihood might be defended and bolstered.[76]

The second meaning of justice is used in relation to nature itself—the claims nature makes on humans. Nature has intrinsic value and dignity, and thus rights, that should be respected. Theorists vary in their specification of the parts of nature that have rights (and deserve justice): from all living beings, to sentient beings, to species, to ecosystems, to the whole earth as a living organic system.[77] Nevertheless, many agree that, in general, the requirements of justice, historically interpreted as giving to each person his or her own due, should be extended to nature.

This position is hinted at in the ecological restoration movement by those who characterize ecosystems (which include humans) as evolutionary ecological entities that have a right to operate according to their spontaneous, self-organizing, and life-generating ways. In this view, the thwarting of nature's autonomy, or its capacity for internal self-renewal and evolutionary ecological adaptation, represents a form of moral injustice. Restoration, therefore, can constitute a form of moral restitution, a giving back to land and its wild others. So although it is humans that are responsible to give justice—to human and nonhuman others—claims of justice, according to this perspective, are lodged both in human and nonhuman beings, and thus both deserve justice.

The notion of justice, in the two ways just cited, has been neglected in developing a meaning and criteria for ecological restoration. Underlining this lack of attention is one of the purposes of chapter 4. As I argue there, raising considerations of the notion of justice in relation to restoration raises the following questions: What voices are not at the table and why? Are there ways in which decision-making processes and procedures are developed and enacted that marginalize certain individuals and groups? Who decides in the first place who will participate, and according to what criteria? And how and why is this or that ecosystem or particular area of land chosen for restoration in the first place? To these ends, careful and sustained attention to the *process* of and *procedures* related to assisting the recovery of ecosystems is required when justice concerns are considered.

Fourth, and finally, consideration of religion-related and ethical dimensions in relation to ecological restoration raises questions of the ambiguous relationship between scientific and religious, cultural and spiritual, secular and sacred understandings of nature and the human relationship to nature. Practical and experience-based understandings of nature's value, for instance, have traditionally played an important role within ecological restoration practice. Further, many restorationists view the fusion of scientific and practical knowledge within restoration practice as one of its most significant contributions to the conservation movement.

But how should personal testimony based on experience that utilizes spiritual or religious language be considered within restoration efforts? And what about creative knowledge and emotional experiences that may be derived from nature-oriented spiritual practices and ritual? How should these be considered within restoration thought and practice, if at all? Questions of this nature surface in the chapters that follow. In some instances, convergence of scientific and religious understandings of ecological restoration may be possible; in other instances, divergence may occur. We can hope that investigation of restoration as not only a scientific but also an ethical and a spiritual practice will create avenues for mutual exchange within the restoration and religious environmental movements.

One of the key premises of this book is that religious approaches to environmental ethics, in this case Christian ethics, can help identify such questions, evaluate the issues they raise, and contribute positively to the development of a religious ethical perspective on restoration's meaning. To potentially make these positive contributions, however, Christian ethical inquiry will also need to be shaped by scientific and experiential understandings of restoration ecological activities.

Thus, we turn next to examine some of the ways in which ecological restoration might do this, particularly in relation to Christian environmental ethical thought.

CHRISTIANITY, ENVIRONMENTAL ETHICS, AND ECOLOGICAL RESTORATION

Christian approaches to environmental ethics, as with environmental ethics more generally as we have seen, are not monolithic with fully agreed-upon methods for ethical inquiry or goals for right or good action in relation to nature.[78] Moreover, different theological, philosophical, sociocultural, political, and economic perspectives and emphases give shape to varying environmental ethical frameworks within the Christian tradition. For example, Ivone Gebara utilizes the contemporary experience of poor Brazilian women in tandem with evolutionary–ecological understandings to develop a cosmological ecofeminism; Sallie McFague relies upon feminist philosophy and liberation theology to construct a liberationist environmental care ethic; and Larry Rasmussen privileges sustainability studies, social ecology, and insights from the environmental justice movement to propose a sustainable communities ethic.

Having said this, however, there are some points of agreement among Christian environmental ethicists. Most basically perhaps is the belief that individuals should be *against* the unjustified destruction of nature and *for* the care of God's earth. Further, and more substantially, is the generally agreed-upon view that Christian environmental ethics should promote the flourishing of all earth's creatures and communities, human and nonhuman together. Although scholars may disagree over precisely which parts or wholes of the natural world ought to be cared for, and according to which reasons, the fundamental idea that earth is sacred in some sense and therefore worthy of reverent respect and care pervades Christian environmental thought. Moreover, Christian environmental ethicists tend to emphasize the notion that human beings have been created with a capacity for living in mutual, fulfilling relation with others, including other creatures and creation as a whole. They also tend to share the conviction that certain classes of people, and their ecologies, have been treated unjustly by more privileged or powerful groups in society and that the healing of earth will also require the healing of relationships between human beings.

With this terrain of Christian environmental thought in mind, let me underline three main contributions that ecological restoration may offer. First, ecological restoration and its practices can help ground Christian metaphors of restoration and redemption in evolutionary ecological understandings of nature and humanity. Restoration-oriented metaphors have long been dominant in the Christian tradition. Yet these metaphors have been applied to individual human redemption at the expense of the earth.[79] Earth, in most Christian conceptions of restoration, has largely been viewed either as the stage on which humanity's restoration-salvation history is played out or as the final burnt (and destroyed) sacrifice prior to Christ's final saving action.[80] Moreover, where Christian environmental thought invokes the metaphor of restoration in discussing the ethical imperative of healing a degraded earth, scientific understandings of ecological restoration have not been explicitly utilized or developed.[81] This is also the case for religious environmental ethics more generally with the exception of indigenous traditions, which are inherently interconnected with restoration of native ecosystems through understandings of traditional ecological knowledge (TEK).

Second, and related to the above, ecological restoration can advance the conversation regarding the ethics of sustainability. Christian environmental ethics regularly lifts up sustainability, ecological integrity, and the integrity of creation as orienting concepts, though when it comes to specifying more concretely what these mean in practice, often only generalities are invoked. Insights from ecological restoration can help to make concrete notions such as ecological integrity and ideas like "the flourishing of individuals and communities together with all creation." Additionally, there has been a lack of discussion within Christian environmental ethics of practices and activities that might enable the societal *transition* toward ecological health and sustainability.

Note that throughout the book I tend to rely on the term "ecological health" or "land health" over "sustainability," using it in the broad sense that Aldo Leopold used it. Land health for Leopold meant the capacity of an ecosystem or land community (that is, soil, plants, animals including humans, energy, sunlight, water, and so on) for internal self-renewal.

Some have critiqued the concept of ecological health as too anthropomorphic, difficult to measure, or flat. As environmental philosopher Holmes Rolston asked me once, is health really all we're after? Don't we want something more than this, in other words, thriving, wonderful life? I, on the other hand, agree with Wendell Berry that "health is wholeness" and, furthermore (following Sir Albert Howard), that "the whole problem of health in soil, plant, animal, and man [is] one great subject."[82] In the context of this project, which is focused on ecological restoration as the healing of damaged landscapes and the healing of the nature–culture relation, land health or ecological health is among the best concepts to represent the sort of wholeness sought.

Having said this, however, I am not against using the term sustainability. Rather, I view sustainability as an overarching umbrella concept to refer to the ability of ecosystems (including human activities such as the economy) to persist indefinitely into the future. I also view sustainability as a lodestar concept, guiding the next phase of global development. The reason, nevertheless, I prefer to more frequently use Leopold's term land or ecological health is because it is more specific and less open to distortion than is sustainability.

For example, although British Petroleum can unabashedly state in its literature that it is committed to environmental sustainability while simultaneously pumping thousands of gallons of particulate matter from its oil refineries into Lake Michigan and into the poor and predominantly African American communities along the northern lake shore, it would be difficult, even ironic, for the company to claim that it is committed to promoting ecological health. Hence, I use the terms in this way—"land or ecological health" to refer to particular conservation and restoration activities, "sustainability" to refer to the general societal transition and search for such. Ecological restoration efforts are key transitional sustainability-making activities that warrant attention within Christian environmental thought.

Finally, Christian environmental ethical thought, as already noted in relation to religious environmental ethics, has tended to focus on right thinking about *nature*, assuming that this will in itself motivate right or good action. Given the urgency of the social ecological crisis, however, individuals, communities, and institutions are deeply in need of sustained thinking around what constitutes right or good *action and living* in relation to earth. Action-oriented visions, structural alterations, and concrete practices need to be worked out according to the various societal spheres (family, religion, economy, politics, and so on) and according to various ecological scales (e.g., community restoration of tall grass prairie to regional plans that weave together agricultural, economic, recreational activities within landscapes). Ecological restoration offers one such example of action and scale-oriented social ecological activity, providing an antidote to the separation of theory from action that runs deeply through the history of Western ethics, including religious ethics.

THE OUTLINE OF THE CHAPTERS

I do not claim in this book to construct a comprehensive restoration environmental ethic. The ethic developed in the pages that follow is largely limited to a consideration of ecological restoration practice in the West, as well as to a treatment of the Christian tradition in particular. This does not preclude its potential relevance for other contexts and traditions, though it does limit the types of sources that are drawn upon and, in turn, the sets of the questions that these sources help to form. Further, the religious restoration ethic developed in this volume does not claim to cover all the critical ethical issues in ecological restoration or Christian environmental thought. Although the ethic constructed here does attend to the major themes that a more comprehensive environmental ethic will need to address—the themes, for example, of the value of nature, norms of social ecological community, and virtues of human ecological identity—it does not claim to treat these themes exhaustively or definitively.

The more modest task, then, of this book is to develop a guide or a framework for an environmental ethic that is explicitly shaped by ecological restoration thought and practice and the Christian tradition. To this end, chapter 1 clarifies restoration's meaning by critically examining key perspectives on ecological restoration, including scientific, biocultural, bioregional, anthropological, philosophical, and technological understandings. Chapter 2 builds on this meaning for ecological restoration, examining the particular interpretations of nature that are generated through the activities of working to heal damaged ecosystems. Chapter 3 widens this exploration by examining the concrete personal and communal experiences of transformation and renewal that restoration ecological symbolic action can generate.

Based on the meanings and experiences of ecological restoration practice explored in chapters 1–3 (part 1), chapters 4–6 (part 2) examine the specific types of values and norms that restoration activities may form. Chapter 4 begins this exploration by broadening the scope of restoration experience to examine the types of social values that characterize good ecological community. Chapter 5 then examines further questions and proposals regarding restoration's potential to function as a symbolic, religious activity, as well as a practical, scientific one. Finally, chapter 6 argues that society is entering a restoration age—one that will require a broader cultural story, or multiple overlapping stories, grounded in restorative actions.

The book's organization and method attempt to embody the active–reflective rhythm of ecological restoration practice in which human beings and communities may become restored to earth in and through the act of restoring earth. Engaging the challenge of pragmatist-oriented philosophical and religious environmental ethics that ethical theory must be grounded in environmental practice and experience, I begin by critically evaluating the environmental practice of ecological restoration as well as contemporary experiences related to such practices. Following this, I move to consider religious and spiritual thought and practice that may contribute to an ecological restoration spirituality and ethic. Finally, I examine the Christian tradition in particular, paying specific attention to interpretations of restoration and redemption. Throughout, I look to theories of social and environmental justice in developing a restoration environmental ethic.

Overall, I argue that ecological restoration can provide a promising model for the nature-culture relation, one that ought to shape twenty-first-century environmentalism as well as environmental ethics. I further argue that the explicit treatment of ecological restoration as an ethical framework may push the field of religious environmental ethics in a more action-oriented, experience-based direction, deepening our understanding of the way in which particular environmental activities may shape certain spiritual and moral. Focusing specifically on a lived environmental activity such as restoration and the concrete experiences it can yield may also helpfully illuminate resources within the Christian tradition that speak to the human relationship to land. Ultimately, ecological restoration's grounding in both environmental science and practical experience provides a distinctive context for examining the paradoxical relationship between scientific and religious, secular and sacred ways of knowing about the natural world.

NOTES

1. Wendell Berry, *Unsettling of America*.
2. Wallace, *Finding God in the Singing River*, 27.
3. Leiserowitz and Fernandez, "Toward a New Consciousness," 61.
4. Mills, *In Service of the Wild*, 3.
5. Maathai, "A Billion Trees," 5.
6. Light, "Ecological Citizenship," 170. Light focuses on ecological restoration in the United States, though I would add that ecological restoration is also becoming among the most pressing and important priorities on the international environmental agenda.
7. Mills, *In Service of the Wild*, 9–10.
8. Peterson, "Talking the Walk," 45–62.
9. Higgs considers agro-ecosystem restoration as a practice that may or may not fall under the species level, depending on the type of project. *Nature by Design*, 96–101.
10. Palmer, Falk, and Zedler, "Ecological Theory and Restoration Ecology," 3–5.
11. Cabin, *Intelligent Tinkering*, 8.
12. Palmer, Falk, and Zedler, "Ecological Theory and Restoration Ecology," 5.
13. Ibid., 4–5.
14. Ibid.
15. See the *Cambridge Dictionary of Philosophy*, 2nd ed., ed. Robert Audi (Cambridge: Cambridge University Press, 1999), for a good historical overview of epistemology scholarship.
16. See Spencer, "Restoring Earth, Restored to Earth," 415–32. I am indebted to Spencer's article for inspiring this book's title and focus.
17. Ibid., 419.
18. Ibid.
19. Ibid.
20. On this idea, see Stephen Kellert's proposal that ethics is the keystone of religion and science, meaning, fertile meeting ground for collaboratively responding to environmental issues. See Farnham and Kellert, "Building the Bridge," 7.
21. See Mill, "Nature," 372–402; and Emerson, "Nature," 380–401.
22. See Aquinas's *Summa Theologica* I-II. 91–94.
23. On the history and emphases of different religious approaches to environmental issues, see Jenkins, "After Lynn White," 283–309.
24. According to White, the Christian interpretation of the dominion texts in the Hebrew Bible (Genesis 1:26), as well as its transcendent view of God and

otherworldly view of salvation, led to an instrumental, anthropocentric, and disenchanted view of the natural world. This view wedded well, White argues, with the mechanized, static understanding of nature, with man as its superior and active conqueror, which developed through the Middle Ages and sixteenth- and seventeenth-century scientific revolution and Protestant Reformation in Europe. It is this nexus of Western science and technology motivated by a Christian anthropocentric worldview that led to the current environmental crisis, according to White. See White, "Historical Roots," 1203–7.

25. On critiques of Christianity within the environmental movement, see Bron Taylor, *Dark Green Religion*.

26. Sittler, "Called to Unity," 43.

27. See, for example, Hessel and Ruether, *Christianity and Ecology*.

28. Mary Evelyn Tucker and John Grim, "Series Forward," in *Christianity and Ecology*, ed. Hessel and Ruether, xviii.

29. See Swimme and Tucker, *Journey of the Universe*. On this project see www.journeyoftheuniverse.org/bios/ (accessed May 26, 2012).

30. Tucker and Grim quote Lynn White Jr. on this idea. White states in his essay in *Science*: "What people do about their ecology depends on what they think about themselves in relation to things around them. Human ecology is deeply conditioned by beliefs about our nature and destiny—that is, by religion." Tucker and Grim, xvi.

31. Ibid, xx.

32. Ibid.

33. Ibid.

34. Ibid., xviii–xix.

35. See, for instance, the work of Bron Taylor, especially his *Dark Green Religion*.

36. See Bron Taylor, "Introduction," *Encyclopedia of Religion and Nature*, 1375.

37. The global religious traditions, or segments of traditions, are not precluded from Taylor's understanding of nature-based religion. Though he is skeptical of Tucker and Grim's claim that the world's religions are in fact directly contributing to the emergence of an environmental ethic on a global scale.

38. On the distinction between the methods of the worldviews' approach and Taylor's lived nature–religion approach, see Bron Taylor, "Introduction," 1377.

39. See, for example, the *Journal of the American Academy of Religion* special issue, "Aquatic Nature Religion," 4.

40. Still other scholars who consider a religious perspective on environmental ethics draw on the evolutionary nature and experience of human beings themselves, suggesting that spiritual and moral values are inherent to the human relationship to the natural world. I draw on this perspective in this book as well. See, for example, the work of Stephen R. Kellert, beginning with his *Kinship to Mastery*.

41. I use the term "environmental ethics" to include both philosophical and religious approaches, though where a distinction is made, I use the terms "philosophical environmental ethics" and "religious environmental ethics," as well as "environmental philosophy" and "environmental theology." Throughout the book, I use

the terms "restoration ethic," "restoration environmental ethic," "environmental restoration ethic," and "Christian restoration environmental ethic" to refer to an environmental ethic developed specifically with restorative activities such as ecological restoration practice in mind. My approach to environmental ethics in this book may fall also under the heading religion and ecology/nature. I use the category ecology/nature because the two main schools of thought within religious studies that deal with environmental issues refer to themselves as scholars of religion and ecology (Mary Evelyn Tucker and John Grim) and religion and nature (Bron Taylor). I view the fields of religious environmental ethics and Christian environmental ethics, although distinct, as subsets both of the fields of religion and nature/ecology and of environmental ethics. On the history and emphases of the thought of Tucker and Grim and Taylor, see Jenkins, "After Lynn White."

42. See, for example, Peterson, "Talking the Walk." See also Jenkins, "After Lynn White," and *Ecologies of Grace.*
43. Peterson, "Talking the Walk," 57.
44. Ibid.
45. See Elliot, "Faking Nature," 81–93; and Katz, "The Big Lie," 231–41.
46. Light, "Ecological Restoration," 49–70.
47. Rolston, "Restoration," 131.
48. Ibid.
49. See Northcott, "Wilderness, Religion and Ecological Restoration," 382–99.
50. Ibid., 392.
51. Spencer, "Restoring Earth, Restored to Earth," 432.
52. Ibid., 431.
53. House, *Totem Salmon.*
54. Mills, *In Service of the Wild*, 160.
55. Ibid., 67.
56. Higgs, *Nature by Design*, 185–95.
57. Higgs, "Good Ecological Restoration?" 342.
58. Ibid.
59. Although a botanist by training, Jordan's work draws on ideas from a variety of disciplines including ecology, philosophy, anthropology, literary criticism, ritual studies, and religious studies.
60. I begin by defining the related categories religious and spiritual, and then ethical and moral.
61. Roof, *Generation of Seekers*, 75–76.
62. Gottlieb, *Greener Faith*, 149.
63. Ibid.
64. See Tucker and Grim, "Introduction: The Emerging Alliance," and Bron Taylor, "Focus Introduction," 867. For more on Taylor's view of green religion, see *Dark Green Religion.*
65. Bron Taylor, "Focus Introduction," 867.
66. Ibid.

67. On the essentialism/constructivism debate as it relates to understandings of nature, see, for example, Soule and Lease, eds., *Reinventing Nature?* On this debate as it relates specifically to damaged and restored nature, see chapter 2 in this book.

68. I am indebted to Eric Higgs's argument in *Nature by Design* for this point.

69. Higgs, "Good Ecological Restoration?" 342.

70. Martinez, "Northwestern Coastal Forests," 67.

71. Higgs, "Good Ecological Restoration?" 342.

72. Peterson, "Talking the Walk," 46.

73. See Jenkins, "After Lynn White," 301.

74. On this view of justice in relation to nature and humans, see, for example, Bullard, *Quest for Environmental Justice*; and Shrader-Frechette, *Environmental Justice*.

75. Schlosberg, *Environmental Justice*, 13.

76. See, for example, the webpage of the IPRN, www.ser.org/iprn/default.asp, as well as Boyce, Narain, and Stanton, *Reclaiming Nature*.

77. Various perspectives on the idea that nature has inherent value and thus ought to be treated with respect can be seen, for example, in Sylvan (Routley), "Is There a Need for a New, an Environmental Ethic?"; Peter Singer, "Not for Humans Only"; Tom Regan, "Animal Rights"; Paul W. Taylor, "The Ethics of Respect for Nature"; Holmes Rolston III, "Value in Nature and the Nature of Value," all in Light and Rolston, *Environmental Ethics*.

78. It is worth noting that Christian environmental thought has historically been advanced by theologians (e.g., Steven Bouma-Prediger, Catherine Keller, Sallie McFague, James Nash, Rosemary Radford Ruether, H. Paul Santmire, Mark Wallace) rather than ethicists. Prominent Christian ethicists who have published widely in the area of environmental issues are far fewer—Larry Rasmussen and Michael Northcott represent perhaps the most notable figures.

79. On this point, see Spencer, "Restoring Earth, Restored to Earth," 428.

80. Ibid., 431.

81. See, for example, Ruether, ed., *Women Healing Earth*.

82. See Wendell Berry, *Another Turn of the Crank*, 89–90.

PART I

RESTORING EARTH

❦

"LET THERE BE A TREE"

A Field Guide to Types of Ecological Restoration

Acts of creation are ordinarily reserved for gods and poets, but humbler folk may circumvent this restriction if they know how. To plant a pine, for example, one need be neither god nor poet; one need only own a good shovel. By virtue of this curious loophole in the rules, any clodhopper may say: Let there be a tree—and there will be one.

—Aldo Leopold, *A Sand County Almanac*

Defining ecological restoration is not an easy task given that meanings are numerous and diverse, often varying dramatically from ecosystem to ecosystem and culture to culture. As a vernacular practice, perspectives on restoration shift according to the types of ecosystems (forest, grassland, wetland, river), degradations (deforestation, erosion, toxification, species loss), and repairs (bioreactivation, recontouring of land or waterways, reintroduction of native species, removal of exotics). Additionally, meanings vary depending on understandings of an ecosystem's original or historic condition, the environmental features that are selected for regeneration, and the goals that are determined for a particular restoration project.

Further complicating any definition of restoration is the fact that restoration is not a new phenomenon, broadly understood, but has a long and varied history, reaching back at least as far as there is record of people interacting intentionally to maintain their natural environments by, for example, shifting crops, fallowing land, or managing certain animal and plant species for consumptive, medicinal, or spiritual purposes. Over time perceptions regarding environmental damage and repair, and thus restoration, have changed significantly. As recently as two centuries ago, for example, most people in the West viewed ecosystem damage as intrinsic to nature's "hideous and dying" ways, as eighteenth-century French naturalist Compte de Buffon claimed.[1] According to this view, damaged land resulted from human neglect to improve upon an intrinsically degenerating nature. Alternatively, ecologists today understand ecosystem damage not as inherent to natural systems but as mostly the result of human activities. According to this contemporary scientific view, it is human rather than nonhuman factors that have caused natural processes and functions to become rundown and therefore incapable of internally renewing the ecological system.

This chapter clarifies the meaning of ecological restoration by presenting a field guide to types of restoration. In this field guide I explain six specific perspectives on restoration's meaning: (a) scientific restoration ecology, (b) biocultural views, (c) deep ecological bioregionalism, (d) anthropology/ritual studies, (e) environmental philosophy, and (f) technological approaches. Subsequent chapters will build on the meanings explored here and pursue questions that they generate.

At the outset, however, I want to note several highlights in recent environmental restoration history, for these have influenced the trajectories of a developing ecological restoration program that now takes the forms I analyze here.[2] Cultures have had and continue to have their own history related to restoring, healing, and renewing land: Environmental historian Marcus Hall charts the divergence, for example, of nineteenth-century European and North American perspectives of restoration based on differing cultural perceptions of ideal and damaged land. On the one hand, "maintaining the garden" as an intensively managed cultural landscape was the prominent restoration view in nineteenth-century Italy; in the United States, on the other hand, "naturalizing the degraded" to pristine, unblemished wilderness was the restoration ideal.[3]

This said, organizations such as the International Society of Ecological Restoration (SER) point to the potential for cross-cultural meanings for ecological restoration, as I show below. The SER emphasizes the environmental historical perspective of the modern-day ecological restoration movement, which originated within North America, and the restoration thought of Aldo Leopold. Given Leopold's significance in relation to broader restoration thought and practice, as well as in relation to this volume, I focus briefly on his work here.

ALDO LEOPOLD THE RESTORATIONIST

As father of the modern-day conservation movement and founder of the arboretum at the University of Wisconsin in Madison (considered the first and most successful ongoing restoration effort in the country), Leopold is thought to be a key progenitor of the scientific ecological study of restoration, a basis for most ecological restoration efforts today. It would be remiss, nevertheless, to suggest that the contemporary ecological restoration movement began with Leopold. Indigenous peoples on this continent (just as in various parts of the world) were in many respects the first restorationists.[4] This is despite the fact that traditional indigenous restoration practices have historically been ignored or minimized within Western approaches to scientific environmental management and restoration techniques.[5] More recently, however, the positive value of traditional ecological knowledge (TEK) applications in ecological restoration efforts has been recognized by many conservation agencies worldwide. To this end, traditional ecological restoration practices are increasingly being used hand in hand with modern scientific ecological techniques.

Within environmental history, restoration began as an aesthetic artistic enterprise in the field of landscape design. During the early part of the twentieth century, designers such as Jens Jensen, Wilhelm Miller, Frederick Law Olmsted, Elsa Rehmann, and Ossian Simmons began creating landscapes modeled on historical landscapes such as the tall grass prairies of the midwestern United States. Prior to this prairies had been little appreciated in the American landscape aesthetic. In this way restoration became a means for "discovering the beauty of a historic landscape, or of ecosystems or processes that had been considered ugly or repellant."[6]

Whereas early-twentieth-century restoration efforts were primarily artistic-oriented ventures, restoration based on the scientific study of ecology is generally understood to have begun with Leopold and his colleagues' experiments at the University of Wisconsin's arboretum in Madison in the early 1930s. After decades of intensive agriculture in the midwest, historical prairie and oak savanna ecosystems, as well as species native to Wisconsin, were virtually nonexistent. In light of this, Leopold and his team (which included early conservation leaders Norman Fassett and Theodore Sperry) proposed developing the Wisconsin arboretum into "a reconstruction of original Wisconsin, rather than into a 'collection' of imported trees."[7] In his speech at the arboretum's dedication during the height of the Great Depression in 1934, Leopold declared: "This Arboretum may be regarded as a place where, in the course of time, we will build up an exhibit of what was, as well as an exhibit of what ought to be."[8]

Leopold was also engaged in his own personal restoration efforts at his derelict farm an hour's drive outside of Madison in Sand County, Wisconsin. He purchased the farm "for its lack of goodness and its lack of highway; indeed my whole neighborhood lies in the backwash of the River Progress."[9] It was this experiment—working to restore the degraded parcel of Wisconsin farmland, home to the Leopold family's retreat-shack (formerly a chicken coop on the property)— that helped to shape Leopold's now-classic book, *A Sand County Almanac*. Each spring the Leopold family of seven would plant thousands of white, red, and jack pines and to a lesser extent tamarack and sumac, working over time to restore more than 150 acres of woodland, prairie, savanna, and wetland marshes.[10]

Among the most important lessons gleaned from Leopold's view of restoration is that conservation work, including restoration, requires tending to the objective task of land health (or pathology) and the subjective task of developing a relationship with the land.[11] Leopold's eldest daughter, Nina Bradley Leopold, for example, quotes one of her father's undated papers: "There are two things that interest me: the relationship of people to each other, and the relation of people to the land."[12] "And as I think about it," Nina Leopold reflects, "both of these things were the main elements of the Shack."[13]

Leopold's contribution to ecological restoration thought and practice has been central to its development.[14] Additionally, many contemporary restorationists point to the historical importance of the founding of the SER in 1988. Whereas other organizations such as the American Society for Surface Mine Reclamation and the Canadian Land Reclamation Association support variations of restoration such as reclamation, restorative repair, rehabilitation, and revegetation, the SER is concerned primarily with holistic restoration, or the restoration of whole ecosystems. The publication of the practitioner journal (really, a cut-and-paste newsletter at first) *Restoration and Management Notes* in 1983 by William Jordan (then a staff member at the Wisconsin arboretum) also helped to bolster the fledgling restoration movement early on. Originally a small organization based in the United States, the SER has burgeoned into an internationally active society with members from more than thirty countries. With Leopold and the SER noted, I now turn to explore the key perspectives on ecological restoration that have developed over the past several decades since the initiation of the scientific study of restoration ecology. I begin with perhaps the most prominent one, a scientific ecological understanding of restoration practice.

SCIENTIFIC ECOLOGICAL UNDERSTANDINGS OF RESTORATION

Although meanings of ecological restoration vary historically, this does not mean that understandings of restoration are completely culturally relative; cultural relativism is not the only alternative to absolute so-called truths in relation to ecological restoration. Rather, as is evidenced in the definitions of ecological restoration that have been developed over the years by the SER, as well as in many restoration projects themselves, it is possible to develop processes of "tested normativity" whereby engaged persons and communities reflect upon, evaluate, and debate the ecological, cultural, and ethical principles that should shape ecological restoration in a particular context.[15]

The SER definition of ecological restoration holds particular promise for achieving a certain level of shared, cross-cultural understanding given that it is communally developed, reflected upon, and tested in consultation with a wide array of international scholars and practitioners working in the field of ecological restoration. Although I do not think the current SER definition goes far enough in terms of including certain cultural and ethical dimensions, I nonetheless view it as an adequate definition to which other perspectives can be compared.[16]

Over the years, the SER revised its core definition several times. In part this reflects the nature of ecological restoration in that it is an adaptive work in progress, continually being shaped by newfound understandings and practices that evolve and change with the dynamism of the land itself. Some have viewed the SER's focus on defining ecological restoration as "endless quibbling," but others have viewed it as "a precondition for deciding what constitutes good restoration."[17] As restoration scholar Eric Higgs argues, "Without the ability to distinguish a good project from a bad one, better projects from worse ones, or even restoration projects from those that are not, the ecological restoration movement—in science, professional practice, community volunteer initiatives, and every other dimension—risks losing its strength of purpose."[18]

The first SER definition put forward in 1990, two years after its chartering as an organization, was the most controversial. It stated: "Ecological restoration is the process of intentionally altering a site to establish a defined, indigenous, historic ecosystem. The goal of this process is to emulate the structure, function, diversity and dynamics of the specified ecosystem."[19] Controversy mostly revolved around use of the term "indigenous," given that both European and Native American restorationists found fault with the view then prevalent among white North American restorationists that it is possible and desirable to restore ecosystems to a precontact or presettlement (indigenous) state.

What would it mean, critics wondered, to restore a site to its indigenous condition given the fact that indigenous peoples have been intensively restoring and managing ecosystems for thousands of years? Further, in places such as Western Europe, which have had extensively settled cultural landscapes for a much longer time than North America, it is virtually impossible to talk about restoring an ecosystem to its historical condition—unless by historical one means nonhumanly populated or one historical stage in cultural evolution.

After extensive polling and conversations with SER members, a revised definition of ecological restoration was developed by restoration scholars Eric Higgs and Dennis Martinez. This definition of ecological restoration read: "Ecological restoration is the process of assisting the recovery and management of ecological integrity. Ecological integrity includes a critical range of variability in biodiversity, ecological processes and structures, regional and historical context, and sustainable cultural practices."[20]

Here the most radical dimension of the definition was the inclusion of the term "sustainable cultural practices" within the concept of ecological integrity. Higgs, for example, worried that it would be difficult to "distinguish between cultural practices that honor participation, modesty, and humility and those that aim to emblazon human pride, greed, and arrogance on nature."[21] Further, he asked, "How do we ensure that a fixation on human values does not swamp the wisdom of ecologists?"[22] Higgs's own response to these questions is that, despite the fact that there are no easy answers, the "surest way is through examining the lessons that flow from practice."[23] Through this exploration "we can glean a few general lessons,"[24] Higgs suggests, for example, that participation, humility, and reflection are crucial cultural practices in restoration.[25]

The current version of the SER definition was developed in 2002. It states: "Ecological restoration is the process of assisting the recovery of an ecosystem that has been degraded, damaged, or destroyed."[26] This pared-down definition is accompanied by an expanded description of restoration in "The SER International Primer on Ecological Restoration."[27] Two key points in the current definition are of special significance.

First, the concept of ecological integrity was jettisoned, given the abstractness of the term that itself would have required extensive additional definition.[28] Higgs still does favor the metaphor ecological integrity to refer to the primary goal of restoration. Ecological integrity for Higgs indicates wholeness, that is, intact systems. He cites biologist James Kay's definition of ecological integrity in which "integrity is an all-encompassing term for the various features—resiliency, elasticity, stress response, and so on—that allow an ecosystem to adjust to environmental change." Nevertheless, the term "recovery," more specifically "assisted recovery" in relation to ecosystems, is used in the SER definition. According to Higgs, "*recovery* refers to the biogeochemical processes that allow an ecosystem to return to the conditions that prevailed prior to disturbance."[29] In relation to restoration, "an ecosystem has recovered—and is restored—when it contains sufficient biotic and abiotic resources to continue its development without further assistance or subsidy."[30]

Implicit in the notion of recovery is the idea that restoration is a process rather than a product. Ecosystems are dynamic entities that change over time; restoring them is similarly a dynamic, evolving process that develops over time; restoring them is a "process of transition, a continuous coming into being of an ecosystem."[31] Conversely, viewing a restored ecosystem as an end product overlooks the fact that some ecosystems, such as the tall grass prairies of the midwest, require long-term ongoing restorative activities.

This emphasizes the point that restoration ultimately is about human beings attempting to participate with and through natural processes. At times this requires relatively brief and minimal contact with an ecosystem; for example, when soil is bioreactivated to jump-start the seedbed. At other times ongoing, sustained efforts are needed: for example, prairie maintenance through routine prescribed burns. Moreover, many restorationists cling to the idea of process as restoration becomes increasingly professionalized and technically managed by large corporate firms or government agencies (or a combination) whose goals often stress the end product of a restored natural area.

Second, the notion of sustainable cultural practices, as we have seen, was dropped from the 2002 SER definition. Sustainable cultural practices seemed too difficult to define in the first place; even if such a definition were developed, the attention to human values would interfere with scientific ecological thinking, according to those scientists who prefer their ecology straight up. However, the notion of sustainable cultural practices is brought back in the SER Primer. It is referenced in relation to developing countries and indigenous perspectives on restoration in particular. The Primer, for example, states: "Some ecosystems, particularly in developing countries, are still managed by traditional, sustainable cultural practices. Reciprocity exists in these cultural ecosystems between cultural activities and ecological processes, such that human actions reinforce ecosystem health and sustainability."[32]

It continues by specifying the role of indigenous ecological management practices in promoting the ecological and cultural survival of indigenous peoples. Further, it states that the North American emphasis on the restoration of pristine ecosystems is indefensible in large parts of the landscapes of, for instance, Africa, Asia, and Latin America, where land use must be manifestly tied to bolstering human survival.[33]

Two additional key restoration concepts proposed by Higgs are worth mentioning in regard to a scientific ecological understanding of restoration. First, the notion of historical fidelity points to the historical character of ecological restoration. Although fidelity is not a familiar ecological term, Higgs argues that it plays a crucial role in ecological restoration efforts. In ecological parlance, historical fidelity means "loyalty to [an ecosystem's] predisturbance conditions, which may or may not involve exact reproduction."[34] Perfect or even near-perfect historical fidelity may not be possible in many restoration projects. Reasons include lack or inadequacy of historical data, financial or personnel shortages, and unavailability of necessary plant or seed stock. Yet, the notion of historical fidelity also involves the idea that restorationists should remain loyal to the goals they have set for themselves and for ecosystems. Restorationists "act as proxies for ecosystems, doing what [they] think is best given a wide array of possible approaches. There is no way of escaping such human contrivance."[35]

This leads Higgs to propose a second key restoration definitional concept, that of wild design.[36] Whereas ecological integrity and historical fidelity are central concepts among most restorationists, wild design is a more controversial idea within dominant scientific restoration thought, despite the fact that Higgs argues for its inclusion in a definition of ecological restoration. The idea that humans can and should "design" nature in restoring ecosystems is, in fact, "anathema to most restorationists, ecologists, and environmentalists," writes Higgs.[37] Given the extent to which nature has historically been dominated and controlled for human purposes, the notion that humans would attempt to intentionally rewrite "the book of Nature" according to their own desires, values, and interests does not sit well with most environmentalists today.

Wild design in the best sense, nevertheless, "is creative intervention according to common and well-discussed ideas," posits Higgs.[38] "Good design is secured by cultural norms, physical (in this case ecological) realities, and imagination."[39] Utilizing the notion of design in relation to restoration provides a certain level of authenticity in recognizing the fact that restoration necessarily involves human agents working actively with natural processes. As bioregional restorationists point out, restoration is fundamentally about human beings learning to actively participate in and cooperate with nature's inherently wild ways—this takes commitment, creativity, imagination, collective effort, and a good deal of hard work. I raise the issue of wild design here as a final restoration concept not only because I tend to agree with Higgs's use of it, but also because it further opens the door for theoretical and practical considerations regarding the extent to which humans should actively intervene within ecosystems.

BIOCULTURAL UNDERSTANDINGS
OF RESTORATION

As indicated above, some restorationists, particularly those from indigenous and environmental justice perspectives, emphasize the importance of cultural practices and the consideration of human needs in relation to ecosystem recovery efforts.[40] To this end, ecological restoration is fundamentally understood as biocultural restoration. Biocultural restoration emphasizes dual aims for restoration practice—that is, not only the recovery of land, but the relationship among people, communities, and the land. "We want to restore life," writes Dennis Martinez.[41] "We want to restore the living and sacred relationship between people and the earth. We want to restore our spirits as we restore the land. We want to restore our culture, our songs, our myths and stories, and the Indian names for creeks and springs. We want to restore ourselves."[42]

Environmental justice and international development scholars similarly tend to understand ecological restoration as a biocultural endeavor. As the introduction to an edited volume on environmental justice and ecological restoration states, "Reclaiming nature as our home means seeing humans as part of nature, not apart from nature. It means rejecting the notion that human beings are like a cancer on the face of the planet, a malignant growth that threatens to destroy its host, and instead recognizing that we can improve our home or degrade it. It means that the task of environmentalism is not merely to contain human impacts within nature's self-healing capacities, but also to enhance nature's capacity to sustain human welfare."[43]

There are a number of ways in which biocultural restorationists understand cultural practices and ecological processes within restoration efforts as mutually reinforcing.

First, in relation to indigenous traditions in particular, culture and nature are viewed as metaphysically integrated spheres. As the IPRN's 1995 founding mission statement suggests, "Indigenous peoples bear a cultural and spiritual tradition that integrates culture and nature. Although this tradition has been badly fragmented under the impacts of modern industrial civilization, it persists to some degree in most traditional communities and has been maintained largely intact in remote places scattered throughout the world."[44]

Second, biocultural restoration understands cultural practices and ecological processes as interconnected through the traditional ecological knowledge systems (TEK) of indigenous tribes and native communities. "By virtue of their survival, all traditional ecological practices are based on self-interest: they are considered life-enhancing and thus sacred to Native peoples," states the IPRN.[45] Additionally, such practices are viewed as sacred in indigenous biocultural thought. Despite the fact that there is great diversity in TEK practices among communities, some basic TEK activities, such as watershed management systems of coastal and island peoples and shifting cultivation in tropical ecosystems, arc shared among various cultures.[46] Critical components of TEK and restoration include property rights and land tenure, responsibility and incentives. Additionally, "community-based decision-making and rules governing resource management also play a large part in preserving biodiversity and encouraging sustainable use" in TEK practices.[47]

Finally, Martinez cites the connection between cultural practices and ecological processes in restoration as reflected in Native American world-renewal ceremonies.[48] The California world-renewal tradition, for example, is crucial to restoration efforts among the Sinkyone people of redwood country. As Julian Lang of the Klamath Tribe describes the tradition, "Our knowledge stems from a race of Spirit Beings whose job *at the beginning of time* was to unravel the mystery, to discover the ideal way of life on the newly created Earth." Lang continues, "In exchange [for the mysteries unlocked and knowledge passed on by the Spirit Beings] we assumed responsibility to Fix the Earth each year: to make the world over, through ceremonial re-enactment of the Spirit Beings' first successful World Renewal. . . . For us the vision of the future is grounded in the responsibility of annually fixing the world."[49]

Biocultural restoration is not only possible or desirable among indigenous peoples and native communities, according to Martinez. It also requires strategies of cultural cooperation. The success of many restoration projects in North America (and perhaps elsewhere as well), writes Martinez, "will depend ultimately on cultures working together and on the European culture now dominant in the [particular] area recognizing the indigenous Indian cultures as part of its own heritage."[50] "What makes ecological restoration especially inspiring," Martinez believes, "is that cultural practices and ecological processes can be mutually reinforcing."[51] "We believe that as a community of ecologists living in times of unprecedented ecological change, we can no longer afford the questionable luxury of working solely within our own traditions if we are to learn to live sustainably. Conserving our options means, in part, conserving the diversity of ways of thinking about problems."[52]

DEEP ECOLOGICAL–BIOREGIONAL UNDERSTANDINGS OF RESTORATION

Ecological restorationists who claim a bioregional social philosophy and deep ecological spirituality advocate a particular type of biocultural restoration.[53] According to bioregional-deep restorationists, restoration may not only assist the recovery of damaged ecosystems but also help in the recovery of modern human beings' recognition of their need to learn how to reinhabit their particular landed places (bioregions). Among the most significant bioregionalists who write specifically on the twenty-first-century promise of ecological restoration as a paradigm for human living within earth are long-time bioregional activists Stephanie Mills and Freeman House. Mills's and House's ecological activist experience goes back to the days of the early radical environmentalist beatniks of the 1960s and '70s; both are eloquent narrators of restoration's promise and possibilities.

Mills summarizes well the bioregional understanding of ecological restoration. In particular, she describes the differences between a bioregional view of restoration as reinhabitation and the dominant view of ecological restoration. The latter implies that human access to the recovering ecosystem is restricted, whereas bioregional reinhabitation involves living and having an economic stake in the land being restored. Reinhabitation, according to Mills, involves deriving "natural provision" from one's life place: "free (but not easy) protein, fuel, and building material."[54] Differentiated from standard ideas of restoration, reinhabitation as a way of life poses alternatives to the present economic system that has necessitated restoration in the first place. Reinhabitation means "the articulation of a culture of place, and for this to come about in our time means the restoration of human community in a society whose members have wildly differing and fiercely held ideas about what land is for."[55]

Bioregional restoration has a dual agenda—both ecological restoration and reinhabitation—but it is reinhabitation that shapes restoration. Given the importance of reinhabitation in the bioregional version of restoration, it bears further explanation here. The term "reinhabitation" was coined by the masterminds of bioregionalism, Peter Berg and Raymond Dasmann. They write: "Reinhabitation means learning to live-in-place in an area that has been disrupted land injured through past exploitation. It involves becoming aware of the particular ecological relationships that operate within and around it. It means understanding activities and evolving social behavior that will enrich the life of that place, restore its life-supporting systems, and establish an ecologically and socially sustainable pattern of existence within it. Simply stated, it involves becoming fully alive in and with a place."[56]

Freeman House has been working out the concept of reinhabitation for the past two decades from his home in the remote region of the Mattole River valley of northwestern California.[57] In the late 1970s a state fisheries biologist told residents that the race of native Chinook salmon (*oncorhynchus tshawytscha*) that ran in the Mattole River was near to extinction. In response, a small group of residents decided to not "stand by and watch while one more race of salmon disappear[ed], especially the one in the river that [ran] through their lives."[58] This small group of residents became "a cohort of several dozen residents [now hundreds] who are spending a great deal of time trying to forge a new sort of relationship to the living processes of their home place."[59]

Forging this new relationship with the Mattole River valley is fully hands-on political environmental activism. Bureaucracies must be dealt with, money raised, alliances formed, contracts entered, weirs (fish traps) built, banks revegetated, and strategies invented, tried, and retried. As this suggests, building working, restorative, sustainable relationships with land is just plain hard work. In 1980, for example, House wrote this in his journal:

In order to learn more about the run [of salmon], and to observe the trap [with weirs], we have decided to maintain all-night watches, in shifts. Although enthralled, most of us had wandered off by midnight, the kids asleep and wrapped in blankets in our arms. John Vargo took over the watch, and at about three o'clock in the morning, he decided to take a more active approach. Poised on a plank stretched between the trap and the shore, he dip-netted two jack salmon and transferred them to the trap. Toward dawn, Greg Smith, a rancher from across the river, came down to relieve John. Together, Greg holding onto John's belt while John plunged the dip net, they captured a beautiful female king salmon [to harvest eggs for reproduction purposes] which must weigh more than twenty pounds.[60]

This sort of unrelenting restoration activism is evidenced in many community-based restoration projects, and it is at times met with sharp opposition from residents who strongly disagree with the idea of restoration to begin with. Such popular opposition to restoration was seen, for example, in the famed Chicago Wilderness project, a coalition of some 100 conservation organizations and government agencies working to restore more than one hundred thousand acres of native Illinois ecosystems within the city of Chicago. Controversy erupted in the mid-1990s, fueled by a small group of residents from two neighborhoods in the northwestern side of the city. Forming a group called Alliance to Let Nature Take Its Course (ATLANTIC) they made claims such as: "God made these non-native plants and trees, just as surely as he made the oak and trillium."[61] "This is an egotistical type of ideology that these forest preserves cannot take care of themselves." "4.5 billion years of history can't be wrong."[62]

Bioregional restorationists such as Mills and House, as well as committed restorationists in general, would whole-heartedly agree with the claim that nature can, in a sense, take care of itself and that 4.5 billion years of history cannot be wrong. What they would sharply disagree with, however, is the idea that humans are somehow not integrally part of that nature and natural history and, further, that active human participation in nature's processes and functions is a violation of nature's own ways. Mills and House, based on their own seasoned and profound experiences with restoration activities, claim that restoration is not an inherently egotistical activity: it is not playing God, it is playing human. To be human, bioregional restorationists believe, is to be active participants, givers and takers, reciprocators, loyal dwellers in and with the land.

ANTHROPOLOGICAL/RITUAL STUDIES UNDERSTANDINGS OF RESTORATION

The restoration scholar who has perhaps made the most significant contribution to the ritualizing of restoration is William Jordan. Jordan is the founder of the practitioner journal *Restoration and Management Notes* (now called *Ecological Restoration*) and the author of *The Sunflower Forest: Ecological Restoration and the New Communion with Nature*. Jordan's understanding of restoration as ritual is largely influenced by the work of literary critic Frederick Turner and his theory of ritual and value-generation.

Turner's understanding of ecological restoration and ritual is evidenced in the title of one of his essays in *Restoration and Management Notes*: "Bloody Columbus: Restoration and the Transvaluation of Shame into Beauty."[63] "More than a technology," writes Turner, "restoration of ecosystems and cultures offers a way of transcending the shame of conquest."[64] It does so first by recognizing and accepting the guilt of the ecological and cultural crimes of conquest, and second, by attempting to make restitution for such crimes. Restitution is made by no less than "the full-hearted effort to make ourselves, and each other, and the future of life on this planet, as beautiful and splendid and generous and creative as they can be."[65] Further, "the classic ecosystems and traditional people we have injured, and to which we, as receivers of stolen goods, are indebted, are dead and cannot be repaid, except by redeeming the promise of the future that they must have been striving for, if they in turn were properly repaying the debt of their own tainted pasts.... Every landscape today—even New Jersey!—is a unique ecosystem that is indebted to the ecological crimes of the past. None of us is clean, and this is no excuse; our very dirtiness ties us to the earth, to life, and claims our commitment to make things better."[66]

Turner proposes that for those of us in the industrialized world, "ecological restoration may be the basis for a world-renewal ritual which, we hope, will be effective in concrete as well as symbolic ways."[67] Jordan echoes Turner's view emphasizing the potential of ritual in restoration to serve as a transformative act for creating, renewing, and revising ethical values. Ecological restoration, according to Jordan, offers a context for the creation of communal rituals, ceremonies, and evocative and performative activities such as gathering and sowing seeds, planting trees, and reintroducing animals in relation to nature. Such acts provide a way to explore and celebrate the contemporary relationship between people and communities and the land.

Further, Jordan writes: "At the deepest level, ritual offers the only means we have of transcending, criticizing, or revising a morality or ethical formulation prescribed by authority or handed down by tradition. Most fundamentally, it is the means by which humans generate, recreate, and renew transcendent values such as community, meaning, beauty, love, and the sacred, on which both ethics and morality depend."[68] The ritual dimension of ecological restoration is especially important for environmentalism, Jordan proposes, because it suggests that reflective action and concrete place-based experience serve as a basis for the creation of meaning and environmental values.[69]

In addition to positive proposals such as Turner's and Jordan's on the potential value of restoration as a new ritual tradition, ritual in restoration has provoked heated debates.[70] Many restorationists, for example, argue that restoration should remain primarily a scientific ecologically based enterprise without any consideration of human values. Others have questioned the spiritual orientation of viewpoints such as Turner's and Jordan's, arguing that humans do not need land-based rituals such as restoration for the salvation of their souls. Still other restorationists have remained ambivalent about the idea of ritual in relation to restoration, focusing instead on the types of civic values that may be generated through the public practice of restoration. It is to this latter perspective I turn next—that is, the viewpoint of many environmental philosophers.

ENVIRONMENTAL PHILOSOPHICAL
UNDERSTANDINGS OF RESTORATION

Within the field of environmental philosophy Andrew Light has been the chief proponent of ecological restoration. Light developed his constructive view of restoration in the 1990s largely in response to the early anti-restoration perspectives of Robert Elliot and Eric Katz. More significantly, however, Light has used ecological restoration as a vehicle for developing an environmental philosophical methodology that is grounded in practice and oriented by the goal of guiding policymakers and citizens on environmental issues. Here we note two key points in Light's environmental pragmatist perspective regarding ecological restoration.

First, Light, in his seminal paper on restoration argues (contra Elliot and Katz) that positive value is created through many, though not all, restoration projects. To make this case Light starts by making a distinction between malicious restorations and benevolent restorations.[71] Restorations such as the type described by Elliot's restoration thesis, Light calls malicious restorations. Those that are "undertaken to remedy a past harm done to nature, although not offered as a justification for harming nature," Light calls benevolent restorations.[72] Light concedes that "we may be able to grant much of Elliot's claim that restored nature is not original nature while still not denying that there is some kind of positive value to the act of ecological restoration in many cases."[73]

In response to Eric Katz's assertion that restoration represents humanity's ongoing domination over nature, Light argues quite the opposite, that restoration can help nature become free of exactly the kind of domination Katz is worried about. Restorations are not, for the most part, dominating technological manipulations imposed on nature by an arrogant humanity, as Katz assumes. Rather, restorations in many cases, Light argues, are encouraging nature to pursue its own course by liberating it from prior human-induced trauma. For instance, many restorations work only to bioactivate soil that has become toxified by hazardous industrial waste.[74]

Ecological restoration has potentially positive value, according to Light, not only because it offers a practice for restoring ecosystem processes (natural value) that have been damaged by human interference and because it offers a practice for restoring the human relationship with nature (moral value) in hands-on, vernacular ways. Restoration activities create moral value, for example, by promoting citizen participation in public environmental practices. Participation in restorations, along with contributing to the good of the community, also can "stimulate the development of moral norms more supportive of environmental sustainability in general."[75] The creation of ecologically oriented, community-based moral norms has the potential to contribute positively to the development of "a new and more expansive 'culture of nature.'"[76]

Light importantly characterizes a culture of nature in terms of the notion ecological citizenship.[77] This is our second point regarding Light's view of restoration. Building on the above understanding that moral value is importantly created in restoration efforts through public participation, Light argues that good citizen political action is characterized by a democratic model of participation (ecological citizenship). According to Light, a democratic model of participation "is the best model for achieving the full potential of restoration in moral and political terms." He continues: "Our choices of how we shape practices and policies involving restoration is a critical test for how deep a commitment to encouraging democratic values we have in publicly accessible environmental practices."[78] The type of democratic ecological citizenship Light has in mind roughly follows "'classical republican' lines ... which identify a range of obligations that people have to each other for the sake of the larger community in which they live."[79] Such might be called an ethical citizenship or a concept of citizenship as vocation, "where being a good citizen is conceived as a virtue met by active participation at some level of public affairs."

Light expands this notion of citizenship to include an environmental dimension. He writes: "If the point of ethical citizenship is to encourage people to take on responsibilities for each other in communities then these responsibilities can be expanded to include environmental dimensions as well."[80] Volunteer restorationists act as good ecological citizens when they participate in the positive public-communal activities of restoration projects. Where restoration projects are conducted by professional design and engineering firms without using citizen volunteers, "an opportunity to foster such ecological citizenship would have been lost."[81] "When people participate in a volunteer restoration, they are doing something good for their community both by helping to deliver an ecosystem service and also by helping to pull together the civic fabric of their home," Light argues.[82] Similar to laws that currently exist to encourage voting, Light proposes that laws should be made that support volunteer participation in publicly funded restoration projects.[83]

TECHNOLOGICAL UNDERSTANDINGS OF RESTORATION

The final perspective on ecological restoration that we note emphasizes a technological approach to ecosystem regeneration. This perspective is most clearly represented in the approach of firms that specialize in large-scale restoration engineering and landscape design projects, often to satisfy corporate objectives. Typically relying on a cadre of experts and professionals in the field, technologically oriented restoration emphasizes the importance of perfecting particular restoration techniques (e.g., plant installation) and efficiently implementing them.

Different from low-tech approaches where restorationists often perform necessary tasks such as digging, planting, and weeding by hand, technologically oriented restoration projects tend to utilize heavy equipment for their application. As one firm states, its work is "what Mother Nature would do if she had a budget, a deadline, and heavy equipment."[84] This same firm advertizes "a highly specialized capability for installing native landscapes. Year-round contracting professionals, augmented by summer field crews, offer complete services in: mechanical or broadcast seeding, mechanical or manual installation of plants, complete site preparation, invasive brush removal, prescribed burning and mowing, licensed herbicide application, erosion control techniques"[85]

Some of these design and technical management firms do the implementation work themselves, but many of the larger firms, such as the one just noted, subcontract the installation, monitoring, planting, and construction work to local contractors who specialize in local systems and plant knowledge, given that much of the technical work of ecological restoration focuses on local plant ecotypes that evolve to a site or region-specific genotype, soil types, and hydrologic conditions. Additionally, many restoration firms run their own highly specialized, large-scale nurseries that grow plants specifically for their projects. The firm cited above, for example, owns and operates a four hundred–acre nursery and farm that produces more than five hundred native species for its prairie, savanna, and wetland restoration projects.[86]

Projects in technologically oriented restoration are often viewed as final, deliverable products, where professional restorationists are hired to install a certain type of landscape according to meet specific objectives (e.g., reduce lawn care costs, improve employee satisfaction and productivity at work). Caterpillar Inc., based in central Illinois, for example, hired a restoration firm to convert its traditional lawn and landscaping to native prairie.[87] In converting lawn areas to native Buffalo Grass lawns, Caterpillar eliminated the need for irrigation and reduced its mowing burden significantly. Prior to the restoration, Caterpillar was spending more than $5,000 per acre on landscape maintenance throughout the campus; after the restoration, costs to maintain the native prairie landscapes are approximately $1,650 per acre. The success of the restoration project, in other words, was measured by its ability to achieve a specific outcome-based goal and objective, namely, to establish low-maintenance prairie ecosystem for economic purposes.

It must be noted that most restoration projects involve setting specific goals and objectives, measurable outcomes, and evaluation plans. Technically focused restoration, however, presents a distinct view of restoration in relation to those that have thus far been highlighted. Namely, it emphasizes technical values, expert knowledge, and outcome-based objectives to the exclusion of other types of values (e.g., social and moral), knowledge (e.g., experiential), and objectives (e.g., inclusion of public participation) that restoration practice may involve.

This type of restoration raises the issue of technological drift—where technical and scientific considerations trump others—within the ecological restoration movement, a trajectory that I critique throughout this book.[88]

I say more about this type of restoration in the following chapter. Suffice it to say here, the technological perspective just described is, to a certain extent, positive—restored prairie is, after all, better, ecologically and economically, than mowed lawn—if ultimately inadequate in terms of defining a meaning for ecological restoration. Further, as I have said, technological types of restoration projects are prominent within the restoration movement. For both of these reasons it warrants inclusion in a field guide to ecological restoration, as well as critical consideration in chapters to come.

CONCLUSION

The six perspectives on ecological restoration just described will each contribute, to varying degrees and in various ways, to the argument developed in the pages that follow. I construe ecological restoration pluralistically according to different perspectives and emphases in order that it might shape a broad and inclusive restoration ethic. The SER definition of restoration will serve as a basic, ecologically oriented definition with potential for cross-cultural agreement. Yet I include additional major perspectives on ecological restoration as well, emphasizing the ways in which these perspectives argue for the inclusion of cultural and ethical dimensions in an understanding of ecological restoration.

The above discussion provides only an initial and a partial exploration of the types of perspectives represented in the restoration movement. Further, additional issues and questions are raised, as noted in the introduction, when religious and ethical dimensions are explicitly considered in a meaning for ecological restoration. Ultimately, these religion-related and ethical issues and questions will need to be addressed if ecological restoration is to provide a promising twenty-first-century model for the nature–culture relation. Without considering these claims, ecological restoration thought and practice will be unable, finally, to provide the restored human-nature connection it purports to promote. Additional chapters will be required to explore the issues that religious and ethical perspectives may illuminate. In the next chapter I begin the constructive task of building a religious restoration ethic, one, as I argue, that is shaped by meanings of nature that are generated through the direct actions of healing damaged land.

NOTES

1. Glacken, *Traces on the Rhodian Shore*, 668.
2. For a volume-length history of ecological restoration, see Jordan and Lubick, *Making Nature Whole*.
3. See Hall, "Co-Workers with Nature," 173–78.
4. On this point, see Martinez, "Northwestern Coastal Forests," 64. Martinez, as already cited, is the cofounder of the Indigenous Peoples' Restoration Network (IPRN), a subset of the largest professional organization of ecological restoration, the Society for Ecological Restoration (SER) International.
5. The history of indigenous restoration knowledge and practices is generally ignored in the dominant environmental historical accounts of restoration scholars. For example, where William Jordan discusses the "pioneers of restoration" he begins with twenty-first-century European-American pioneers rather than considering indigenous Americans as the first real pioneers of restoration. See Jordan, *The Sunflower Forest*, 87.
6. Ibid., 86.
7. Higgs, *Nature by Design*, 78.
8. Ibid.
9. Leopold, *Sand County Almanac*, 46–47.
10. Mills, *In Service of the Wild*, 93–94.
11. Ibid., 96.
12. Ibid., 110.
13. Ibid.
14. On this point and the historical markers noted here, see Higgs, *Nature by Design*, 78–82.
15. See Peters, *In Search of the Good Life*, for more on the ethical notion of tested normativity, 19.
16. It is important to note that the other perspectives on ecological restoration discussed below are not in disagreement with the SER definition; in fact, most of the authors used in this section are members or supporters of the SER and its views. Nonetheless, the various viewpoints presented in this chapter warrant particular attention because they offer significantly different and additional emphases to the SER definition. Further, laying the various perspectives on ecological restoration next to one another allows a fuller picture of restoration to emerge in terms of its potential impact on contemporary environmentalism.
17. Higgs, *Nature by Design*, 93.
18. Ibid., 96–101.
19. Ibid, 107.
20. Ibid., 109.
21. Ibid.
22. Ibid., 121–22.
23. Ibid.
24. Ibid.

25. Ibid.
26. Science and Policy Working Group, "SER International Primer," 3.
27. Ibid.
28. See Higgs, *Nature by Design*, 122.
29. Ibid.
30. Science and Policy Working Group, "SER International Primer," 3.
31. Higgs, *Nature by Design*, 110–11.
32. Science and Policy Working Group, "SER International Primer," 2.
33. Ibid.
34. Higgs, *Nature by Design*, 127.
35. Ibid., 128.
36. Ibid., 13.
37. Ibid.
38. Ibid.
39. Ibid.
40. There are many overlaps with the biocultural restoration perspective of the IPRN and that of the environmental justice movement. Further, scholars and activists who work in the environmental justice field often work with indigenous communities and communities that rely upon ecosystems for their survival. On this view, see, for example, Boyce, Narain, and Stanton, *Reclaiming Nature*.
41. Martinez, "Northwestern Coastal Forests," 67. Martinez writes this specifically in relation to the biocultural restoration project of the Sinkyone Intertribal Park in northern coastal California, although it represents well the view of the IPRN in general.
42. Ibid.
43. Boyce, Narain, and Stanton, introduction to *Reclaiming Nature*, 3.
44. As already noted, the IPRN is a working group of the SER, developed at SER International's annual conference in 1995, which was held in Seattle, Washington. See Society for Ecological Restoration International, "1995 IPRN Founding Mission Statement," Indigenous Peoples' Restoration Network, www.ser.org/iprn/default.asp (accessed November 1, 2007).
45. Ibid.
46. Ibid.
47. Ibid.
48. Martinez, "Northwestern Coastal Forests," 69.
49. Julian Lang, introduction to *To the American Indian: Reminiscences of a Yurok*, by Lucy Thompson (San Leandro, CA: Heyday Books, 1991), quoted in Martinez, "Northwestern Coastal Forests," 65.
50. Martinez, "Northwestern Coastal Forests," 69.
51. "SER International Primer," 2.
52. Jesse Ford and Dennis Martinez, "Welcome to the Indigenous Peoples' Restoration Network," International Peoples' Restoration Network, www.ser.org/iprn/default.asp (accessed November 1, 2010).

53. On the integral connection between deep ecology and bioregional thought see Bron Taylor, "Deep Ecology," 269–99.
54. Mills, *In Service of the Wild*, 160.
55. Ibid.
56. Berg and Dasmann, "Reinhabiting California," 217–18.
57. See his *Totem Salmon*.
58. House, *Totem Salmon*, 3.
59. Ibid.
60. Ibid., 150.
61. Shore, "Controversy Erupts," 26.
62. Ibid., 35.
63. Turner, "Bloody Columbus," 70–74.
64. Ibid., 70.
65. Ibid.
66. Ibid., 73.
67. Ibid.
68. Jordan, *Sunflower Forest*, 5.
69. Bron Taylor, *Encyclopedia of Religion and Nature*, s.v. "Restoration Ecology and Ritual."
70. On this, see Meekison and Higgs, "Rites of Spring," 73–81.
71. Light, "Ecological Restoration," 49–70.
72. Ibid., 54.
73. Ibid.
74. Ibid.
75. Light, "Ecological Citizenship," 176.
76. Ibid.
77. Ibid.
78. Ibid.
79. Ibid.
80. Ibid., 178.
81. Ibid., 182.
82. Ibid.
83. Ibid., 179.
84. "Contracting Brochures," Applied Ecological Services Inc., www.appliedeco.com/Profiles.cfm (accessed May 10, 2012).
85. "Prairie/Savanna Management," Applied Ecological Services Inc., www.appliedeco.com/PrairieSavanna.cfm (accessed May 10, 2012).
86. "Taylor Creek," Applied Ecological Services Inc., www.appliedeco.com/Index.cfm (accessed May 10, 2012).
87. "Customers," Pizzo & Associates Ltd., http://pizzo.info/index.php/customers/caterpillar (accessed May 10, 2012).
88. On the idea of technological drift within restoration, see Higgs, *Nature by Design* and "Good Ecological Restoration?"

FOR THE SAKE OF THE WILD OTHERS

Restoration Meanings for Nature

Every time an ecosystem is restored, a particular view of nature blooms brighter. Hence, restorationists are central agents in the definition and redefinition of what is, and what counts as, nature.

—Eric Higgs, *Nature by Design: People, Natural Process, and Ecological Restoration*

Defining what is and what counts as nature has been one of the main preoc-cupations of contemporary environmental ethicists. Depending on the phi-losopher one converses with, all biological life (Arne Naess), trees (Christopher Stone), sentient animals (Peter Singer), whole ecosystems (J. Baird Callicott and Holmes Rolston), and even larger bioregions (Peter Berg) count as part of nature and have value that makes certain moral claims on humans.[1] Religious environ-mental ethicists too have argued in favor of recognizing a moral status for parts and/or wholes within the nonhuman creation. In this case, however, the natural world and its beings hold intrinsic worth that appeals to moral sensibilities based on nature's relationship to the sacred. Creatures are endowed with the glory of God (Sallie McFague), for instance, or the whole creation is characterized by an integrity that originates and participates in God's very being (Larry Rasmussen).

But what about when we consider the value of nature where land has been damaged and then repaired by human practices? How does this change the assessment of nature's moral standing? What understandings about nature bloom brighter, as Eric Higgs asserts, when land is restored? And why dig in our heels and restore degraded ecosystems in the first place? Why not just let nature take over where it may (or may not), and move onto greener (or browner) pastures?

My reasons for explicitly considering the value and meaning of nature, despite its well-wornness as a philosophical and theological subject, are twofold: First, the notion of nature serves as the orienting concept for restorationists in particular and environmentalists more broadly. Second, and more significantly, experiences, beliefs, and activities in relation to nature take on new meanings in restoration light, for now these phenomena are interpreted from within the medium and activity of healing damaged land. As we saw in the previous chapter, for example, some types of restoration narratives about nature's value border on the religious, despite restoration's scientific basis. Further, in theological tropes, considerations of a wounded and healing earth help concretize well-worn notions such as the integrity and sacredness of creation.

In this chapter I explore key ways in which dominant environmental ethical notions of nature's value and creation's sacredness are recontoured when damaged and healing land in particular is considered. In chapters 5 and 6 I broaden this exploration of the meaning of nature, proposing that restoration thought and practice can contribute significantly to the symbolic re-storying of our relationship to the natural world. Here, I begin by responding to the charge (seen before in the introduction) put forward by restoration's first philosophers, Robert Elliot and Eric Katz, that restoration is actually a form of human domination over an otherwise wild and free nature, for the premise and argument of this book offer a different view of restoration; as I argue, restoration, contra Elliot and Katz, is a positive and constructive movement that can create meaningful and ethical views of nature and the human relation to nature. In further elaborating this point about restoration's value, I consider the issue of deconstructionism and the debate between constructed and essentialist views of nature, proposing that restoration thought, based on the material hands-on activity of working with natural processes, offers a middle path between these extremes. With this middle view of nature in mind, I propose six dimensions in a meaning for nature based on both restoration and religious environmental understandings. In closing I argue that this complex restoration meaning of nature, including the element of its intrinsic-sacred value, can and should be drawn on as a strand in building a broader restoration narrative and ethic.

FOR THE SAKE OF THE WILD OTHERS—DON'T RESTORE!

To a large extent, the philosophical questions that dominated the field of ecological restoration in the 1980s echoed those initiated a decade previously within the broader field of environmental ethics. Does nature have nonanthropocentric, intrinsic value, and if so, which natural features make moral claims on humans? Is it possible, and desirable, for human beings to replicate (as in restoration) or mimic nonhuman nature in any way, and if so, which values ought they follow? If not, what is the proper human stance in relation to wild nature? Are culture and nature fundamentally separate or overlapping categories, or are such distinctions passé in postmodern light?

Questions related to nature's value within restoration thought, however, have not only echoed but also diverged from and advanced beyond those of dominant environmental philosophies. Restoration queries probed, for example: If nature is understood nonanthropocentrically, what actually *happens* to nature's intrinsic value when ecosystems are damaged by human activities? If nature has objective intrinsic value apart from human valuation of it, can human activities such as restoration ever regenerate this type of natural ecological value? If the most natural ecosystems are those that maintain a certain level of their original, intrinsic wildness, is there anything inherently good about restored natural ecosystems?

As we have seen, early restoration philosophers Robert Elliot and Eric Katz did not paint a rosy, or even slightly favorable, picture regarding restoration's potential as an ethical environmental practice. Contrary to what one might expect at first glance—that is, that humans ought to restore natural values where they have been wrongly damaged by humans in the first place—neither Elliot nor Katz used arguments regarding nature's value as ethical justification for the practice of restoring damaged ecosystems; rather, they used such arguments against it.

Elliot based his own antirestoration argument on a particular assumption regarding the source and nature of nature's value. On the one hand he conceded that restoration activities may be capable of rehabilitating certain environmental features, what he calls "objects"; that is, certain landscape formations, habitats, and species can be put back. On the other hand there remains a dimension of natural value, according to Elliot, that is ultimately irreplaceable, impossible for humans to re-create or regenerate: an ecosystem's origins. "There is at least a prima facie case for partially explaining the value of objects in terms of their origins, in terms of the kinds of processes that brought them into being."[2] John Muir, for instance, valued Hetch Hetchy Valley "not just because it was a place of great beauty," writes Elliot, "but because it was also part of the world that had not been shaped by human hand; his valuation was of a literally natural object, of an object with a special kind of continuity with the past."[3]

Elliot's primary contention was that ecological restoration was faking nature, meaning that a restored nature was analogous to a forged piece of artwork; it attempted to pass as the original, despite the fact that it could never be so. For Elliot there is no possible way for any restoration project to (re)create or restore natural value. This is because for something to be natural, according to Elliot, it must be "unmodified by human activity," retaining complete, unadulterated continuity with the original ecosystem. A "regenerated environment does not have the right kind of continuity with [the one] that stood there initially; that continuity has been interfered with by the earlier devastation," writes Elliot.[4] Once a forest's or woodland's or bog's or other ecosystem's natural value has been destroyed, damaged, or degraded, there is no going back; what is lost is lost, according to Elliot, at least where nature is concerned.

Elliot's broader worry was that ecological restoration would become a justification—used primarily by developers and natural resource-extracting firms—for the destruction of nature. "In other words [mining companies, for example] are claiming that the destruction of what has value is compensated for by the later creation (re-creation) of something of equal value."[5] Elliot calls this the restoration thesis, that is, the (false) idea that natural value can be (re)created after it has been destroyed. In other words if we strip mine it, we can put it back, re-create the mountaintop, replant the trees, reintroduce the animals, re-dig the lakes and rivers, or so Elliot feared the rationalizing would go. Restoration would become a fancy (or perhaps brute) economic justification for the massive corporate destruction of nature.

Granted, naturalness can be thought of on a continuum, admitted Elliot. For example, a rehabilitated forest on an old strip mine is more natural than a simulated, plastic, Disney-like wilderness area. "Still, the regenerated environment does not have the right kind of continuity with the forest that stood there initially; that continuity has been interfered with by the earlier devastation."[6] Restoration might be able to put some of nature back—its parts, processes, and functions—but as a cultural activity it can never put back an ecosystem's original value. We might better think of restorations in the same category as art forgeries. According to Elliot, both represent fakes; neither can ever have the value of the original.

A decade after Elliot's challenge environmental philosopher Eric Katz registered his outrage with the restoration idea. Though Katz used Elliot's arguments as a starting point, Katz's interest was to issue a "deeper investigation into the fundamental errors of restoration policy."[7] Restoration policy, Katz argued, "is based on a misperception of natural reality and a misguided understanding of the human place in the natural environment . . . it is the same kind of 'technological fix' that has engendered the environmental crisis."[8] Katz writes that his reaction to the idea of restoration "is almost entirely visceral." "I am outraged," he states, "by the idea that a technologically created 'nature' will be passed off as reality. The human presumption that we are capable of this technological fix demonstrates (once again) the arrogance with which humanity surveys the natural world."[9]

Katz's argument is, like Elliot's, predicated on the view that a restored nature is artifactual, in other words, culturally created and thus in distinct contrast to the natural.[10] *Natural* for Katz is a term that should be used "to designate objects and processes that exist as far as possible from human manipulation and control."[11] Insistently maintaining that there is no possible way for humans to intervene in nature without disrupting nature's spontaneous ways, Katz asserted that a restored nature is an anthropocentrically imposed nature. It is not a nature that is "permitted to be free, to pursue its own independent course of development."[12]

Moreover, there may be an even greater danger in proposing ecological restoration as a conservation strategy, worried Katz in response to Elliot. More than producing a forged, faked nature, as Elliot had argued, restoration represents an extreme, if masked, form of human domination and control over nature. Restorations are nothing but cultural artifacts, a nefarious example of humans once again attempting to manipulate nature according to their own preferences and tastes.

"The big lie" for Katz is that restoration claims to regenerate natural systems. "On a simple level, it is the same kind of 'technological fix' that has engendered the environmental crisis. . . . On a deeper level, it is an expression of an anthropocentric world view, in which human interests shape and redesign a comfortable natural reality . . . on the most fundamental level, it is an unrecognized manifestation of the insidious dream of the human domination of nature."[13] Restored nature, according to Katz, is merely a cultural artifact created by contingent human interests and for human use, according to Katz.

Artifacts, in contradistinction to natural objects, are created for a particular purpose and function, namely human ones. They are essentially anthropocentric. In fact, argued Katz, "it would be impossible to imagine an artifact not designed to meet a human purpose."[14] The meaning of artifacts is predicated on the doctrine of anthropocentrism. Generally this is not problematic, given that humans create and use artifacts primarily in social and cultural contexts. "But once we begin to redesign natural systems and processes, once we begin to create restored natural environments, we impose our anthropocentric purposes on areas that exist outside human society."[15]

For Katz, the natural is to be distinguished from the artifactual because the natural is defined as "independent of the actions of humanity."[16] Therefore, in no way can human restoration regenerate the nature of natural objects. Nineteenth-century utilitarian philosopher John Stuart Mill raised these problematic aspects of defining nature in relation to humanity long ago. Katz recognizes this point as well. Interestingly, Katz wants to maintain a fundamental distinction between nature and culture similar to Mill's, although his solution for how humans should act in regard to nature is opposite from Mill's. In short, Katz argues that humans ought to respect nature's autonomous ways by staying away, whereas Mill proposes that the only proper moral response to nature is to attempt to amend and fix it to bring nature to a higher standard of goodness and justice.[17]

Katz acknowledges that there is virtually no part of the natural world at this point in history that remains completely independent of human influence. He also recognizes that human activities are in a sense natural, given that humans are naturally evolved beings. And humans are capable of engaging in some activities that are more natural than others. Consider natural childbirth, Katz suggests (following Andrew Brennan's use of the example): "It is natural, free, and wild not because it is a nonhuman activity—after all, it is human childbirth—but because it is independent of a certain type of human activity, actions designed to control or to manipulate natural processes."[18]

Truly natural objects and processes are autonomous, in this view, in the sense of being independent from human manipulation or control. Some restorations may be less anthropocentric and manipulating than others (that is, a restored self-sustaining forest is better than a tree plantation). Still, for Katz, all restoration efforts involve "the creation of artifactual natural realities, the imposition of anthropocentric interests on the processes and interests of nature."[19] Restoration activities disallow nature from pursuing its own independent course of development; its fundamental freedom, autonomy, and capacity for self-realization are denied. Domination, "the denial of freedom and autonomy," is restoration's damning error for Katz.[20]

This points to one of the fundamental differences between the view of the nature-human relationship offered by Katz and Elliot, on the one hand, and most restorationists on the other. Katz and Elliot believe nature and culture should remain sharply divided in order to preserve nonhuman wildness; restorationists emphasize the co-evolutionary and co-creative promise of ecological restoration in providing a metaphysical and practical bridge between a (wrongly) divided nature and culture. Elliot's and Katz's wholesale negative views of restoration, however, disallow them engaging the possibility that there may in fact be some kind of positive value created through certain kinds of restorations. Katz's belief that restoration is an arrogant, egotistical idea based on human domination over nature, for example, overlooks alternative understandings and types of restoration. In practice (which neither Elliot nor Katz pays much attention to) it may be that certain kinds of restoration efforts can produce immensely humbling and valuable relationships between human beings and nature, as I argue in this book.

For both Elliot and Katz, nonhuman nature has objective value by virtue of its self-origination in evolutionary biological processes that are fundamentally independent of human agency. Humans are to respect nonhuman nature's inherent capability for self-generation, renewal, and evolutionary historical development by leaving it alone to the greatest possible extent. Some human phenomena such as natural childbirth may follow, submit to, or respect true natural evolutionary processes, but restoration is definitely not one of them for Elliot and Katz. It is best to focus our efforts, conceptually and practically, on attempting to preserve the truly wild natural values that still exist. For the sake of the wild other(s)—don't restore, stay away, let nature be free!

WRITING AND DESIGNING THE LANDSCAPE:
THE CHALLENGE OF DECONSTRUCTION

Postmodern deconstructionist philosophers, theologians, and scientists have questioned positions such as the ones just cited that posit an essential, given reality for nature apart from human valuation of it. Echoing the work of French philosopher Jacques Derrida, deconstructionist environmental authors such as J. Baird Callicott have critiqued the idea of nature itself, and language in general, in terms of its capacity to correspond with the physical world as it is.[21] Alternatively, deconstructionists posit that language, as well as the perception and experience upon which it is based, is historically contingent. Even as philosophers and scientists attempt to present stable descriptions of nature and ecosystems, deconstructionists believe that these descriptions are fundamentally reflections of particular cultural values and norms. The concept of nature, in counterdistinction to culture, is primarily a product of western intellectual thought; it is not a universally applicable description of the way in which the material world is organized. Furthermore, deconstructionists argue that such understandings of nature and culture are products of the same binary logic of thought that is responsible, in part, for creating the environmental crisis in the first place. What we need instead, they suggest, are meanings for nature that are plural, overlapping, emergent, and contextual.

Essentialist-oriented environmental authors have presented arguments regarding nature's meaning contrary to the ones above.[22] For them, nature operates, in some sense, according to certain objective universal (essential) processes and functions that can accordingly be observed and described empirically. According to this position, the idea of nature may be culturally contingent to a certain degree, yet it is also the case that ecosystems can be defined in terms of their constitutive capacities and functions.

Debates over nature's constructed or essential (or a mixture) character have been central to accounts of nature's moral value within restoration thought as well. On the one hand, this reflects a broader trend within academic disciplines over the past two decades to consider postmodern deconstructionist perspectives on the nature and meaning of reality. On the other hand, restoration practice is especially susceptible to questions related to what is, and what counts as, nature, given its character as a human activity attempting to work within and through natural processes. Restorationists are continually faced, for example, with the simultaneous realities that (a) they are in some sense making up nature as they go, based on their own personal experiences and cultural values *and* that (b) natural processes are in some sense other than a projection of the human mind.

The question, as we have already seen, of how best to describe, interpret, and practically and emotionally deal with this dual reality is one that environmentalists, including restorationists, have grappled with extensively and variously over the past twenty-five years. Some restoration theorists, such as Elliot and Katz, have stressed the reality of nonhuman natural beings and processes, independent of ideational de/construction. Others have criticized what they view as an essentialized, reified understanding of nature or ecosystems, emphasizing instead the contingent and values-based character of these concepts. Two examples, one from philosophical ethics and one from scientific ecology, represent deconstructionist proposals within restoration thought.

First, philosopher Alan McQuillan proposes that land managers adopt a postmodern deconstructuralist viewpoint in "defending the ethics of ecological restoration."[23] Poststructuralist understandings of nature, according to McQuillan, are more aligned with empirical scientific understandings—most notably, the evolutionary sciences of chaos theory. Essentialist views such as Elliot's and Katz's posit an objective essence or *noumenon* within nature that is ontologically lost when a natural area is damaged by human activities, argues McQuillan. Yet a *noumenon*, as eighteenth-century philosopher Immanuel Kant argued, can never be known by empirical science, which can only observe and test the phenomenal world. Hence, essences of things "are more properly the provenance of religion than of science," according to McQuillan.[24] Discussions of them ought to be left out of restoration ethical (and especially scientific) thought altogether. In other words, considerations of the "really real," sacred dimension of nature (if there is one) ought to be left to the scholars and practitioners of religion. Although one might balk at McQuillan's conclusion—that consideration of the numinous dimensions of the natural world ought to be relegated to the field of religion—he nonetheless highlights an important point related to postmodern understandings of nature. That is, what McQuillan most wants land managers and restorationists to glean from deconstructionist thought is its skepticism regarding privileged or absolute knowledge about the real. Restorationists should "avoid speaking of originality" in relation to ecosystems, given the way it reflects the idea that there existed a prior ecological state that was objectively, essentially more natural.[25] "As ecologists and restorers we make up 'nature' as we go along, not in contemptuous denial of the real, but in attempted simulation of what we perceive the real to be."[26] Restoration, contra Elliot and Katz, can be "a joyful act" when we recognize its character as a creative act of "writing the landscape."[27] We might think of restoration as akin to gardening or creating art, writes McQuillan. Just as "creativity is not precluded by physical constraints in the plastic medium," McQuillan states, neither should it be in the earthy environmental medium of the restorationist.[28]

From this viewpoint McQuillan observes that postmodern thought does not necessarily omit entirely the notion that nature is "real."[29] For instance, post-Freudian epistemologist Jacques Lacan proposed that the human mind is constructed to register the notion of nature along three lines: the symbolic, the imaginary, and the real. The symbolic is the world of language, everything we can communicate; the imaginary lies in the unconscious, the basis of desire and creativity; and the real is that which is beyond or other than the mental world, the "truly wild" breaking into our minds.[30] "The real is the very stuff of life and death itself, but it is always just beyond cognitive grasp," writes McQuillan. "The closest we can come is when we devise symbolic models, such as nature.[31]

A second example of deconstructionist restoration thought comes from restoration ecologists Mark Davis and Lawrence Slobodkin.[32] They too propose that restorationists should rely upon deconstructionist views of nature, though this time, for primarily scientific ecological reasons. The new ecology, based on the "new sciences" of quantum physics and chaos theory, argues that communities such as ecosystems are not as integrated and coherent as historically assumed. Since the 1930s, with the groundbreaking work of E. F. Clements and his colleagues, ecological thought has assumed, to varying degrees, that ecosystems are stable, tightly organized entities. Yet a contemporary scientific ecological view, according to Davis and Slobodkin, suggests that communities, unlike individual organisms, lack any sort of distinct boundaries or evolutionary imperative. For instance, they have no evolved mechanisms for producing certain processes such as reproduction, growth, or death.[33]

Since "ecological communities and ecosystems lack any intrinsic evolutionary or ecological purpose," write Davis and Slobodkin, "one cannot validly invoke any ecological (or evolutionary) rationale to establish particular restoration goals."[34] Ecosystem integrity and health are value-based assessments, not scientific ones, they argue. There may be species and communities that exist in particular geographic locales, but this does not mean there is some integrated, organism-like entity such as an ecosystem whose life can go better or worse.

Restorationists would do best to think of their work as *ecological architecture*, Davis and Slobodkin propose, a term that recognizes in restoration efforts the role of both science and values.[35] Scientific ecology and evolutionary biology may be utilized in the implementation of restoration objectives, though the identification of restoration goals is "fundamentally a value-based social enterprise."[36] Whether people prefer "a historical environment, a species-rich environment, a particular set of species, or some other type of landscape, restorationists cannot logically or ethically invoke ecology or evolution as a justification for these preferences."[37]

Whereas Katz fears and rejects precisely this idea that restorationists "make up nature as they go along," deconstructionist restorationists such as the ones just noted affirm and celebrate it. If restorationists would only recognize and acknowledge restoration for what it is—that is, an intentional human act based on contingent views of environmental damage and repair—they might better justify it as a distinctive and significant conservation practice, deconstructionists argue. Restored, artifactual nature is not bad, and restorations are not sheer fakes, according to this view. Insofar as restoration is understood as a creative and potentially joyful act of writing or designing the landscape it can be defended on ethical grounds.

The encounter of Christian environmental thought with deconstruction perspectives has moved it to reexamine some of its most valued notions, such as the integrity of creation. The recognition that such concepts reflect attachments to deeper assumptions regarding understandings of reality has given rise to serious questioning of both the assumptions and the accepted interpretations of reality.[38] Just as McQuillan invites restorationists to engage in a healthy skepticism regarding privileged or absolute knowledge of the real, so too some Christian environmentalists now critically consider the ways in which their particular beliefs about nature shape their accounts of reality, the world, God, and creation. The point here is not that claims, scientific or religious, regarding how things really are are wrong or unwelcome. Rather, it is that they require critical evaluation in terms of the ways in which the theorists' or activists' personal values and worldviews are constructive of the picture of reality they give.

In my own case, this project has made me cognizant of the ways in which my Protestant (Dutch Reformed) upbringing, as well as my relative privilege and routine experiences in pristine wilderness settings, have formed a certain lens on the world that favors seeing integration, order, and natural beauty. Additionally, however, my experiences and education living and working in contexts where both people and land are marginalized and treated unjustly have oriented me toward the underside of life, and the human and nonhuman suffering that does not have to be in society and in nature. My interest in the topic of restoration, therefore, with its dual emphasis on nature's historical mistreatment as well as nature's potential for exhibiting astonishing beauty and vitality and its need therefore for restoration comes as no surprise. My encounter with deconstructive philosophy, however, invites me as well as other theorists to reflect critically on the ways in which my experiences and assumptions are both contributing to and implicated in the narrative of reality being told. Deconstruction thought helps to keep it "honest in its reminder that all claims to reality are partial and reformable," as ecological theologian Mark Wallace states.[39] It serves as a healthy tonic, purifying ecological theology, writes Wallace, of its "essentializing tendencies" and readying it for entrance in public environmental discourse. Cognizant that meanings for nature are shaped by cultural and religious assumptions, theology can enter the public domain, Wallace suggests, "clear-headed about its own founding assumptions and clear-sighted in its distinctive vision of an interdependent world charged with the healing power of the Spirit in all things."[40] The same might be said for restoration religious thought in terms of its capacity for contributing to a broader, public restoration ethic. Insofar as it is characterized by a healthy measure of critical reflection regarding its assumptions about a wounded and healing creation, it may be able to enter the public sphere with a distinctive vision of land's and people's regeneration. We return to this idea in chapter 6. For now we continue to develop the proposal that restoration presents a distinctive view of and meaning for nature, based on its orientation toward regenerating damaged natural lands.

A STRATEGIC ESSENTIALIST VIEW OF NATURE

There is a middle path, in my view, between the worlds of deconstructionism and essentialism. On the one hand I agree with environmental philosopher Baird Callicott that "Nature as Other is over.... The modern picture of nature is false and its historical tenure has been pernicious. A new dynamic and systemic post-modern concept of nature, which includes rather than excludes human beings, is presently taking shape."[41] Particularly given the extent to which humans have historically seen themselves as masters over and conquerors of an otherwise amoral, or immoral, disorderly, and irrational nature, an ecological concept of nature that includes human beings and their cultures is necessary. On the other hand, although culture is grounded in and surrounded by nature, it is not entirely bounded by it. Humans today have the capacity to alter their environments and even their own psyches in ways that, in part, transcend their given biological contexts.

The overarching goal of environmental ethics is to create more cooperative and meaningful relationships between humans and land. Sometimes this will require that we stress the fundamental, evolutionary ecological unity between people and nature; sometimes it will necessitate that we emphasize people's unique moral capacities and responsibilities for land's restorative care. Along these lines I propose that a critical constructivism, or strategic essentialism, best reflects our experiences of nature as both continuous with and differentiated from the human mind.[42]

A strategic essentialist view of nature involves a recognition that views of nature are based on personal and cultural perception and that there is a distinct or connected reality beneath or under this perception. Most ecologists today, for example, hold onto a view of ecosystems as (loosely) integrated entities, despite the challenges posed by postmodern deconstructionist thought.[43] Ecosystems can, for example, exhibit resilience in responding to perturbations and dynamism in terms of level of functioning at a given point in time. Ecologists tend to agree that ecosystems can no longer be accurately thought of as evolving pseudo-organisms as they were two or three generations ago. Nevertheless, there remains "a great deal of evidence that biotic communities and ecological systems, at various scales of complexity and resolution, do show a very high degree of integration, or 'coherence' in their responses to perturbations of various kinds."[44] Ecosystems may not reproduce in the evolutionary biological sense, though they do develop dynamically, with emergent properties and feedback loops, over time.

In part we are always making up nature through our ongoing, continually shifting cultural beliefs and practices, and restoration certainly contributes to this shaping, or as Higgs calls it "blooming," of beliefs about nature. Yet humans do not wholly create or make up the "really real" value of ecosystems as we go along. Natural processes have a way of poking through our human constructs and practices, "exposing the hubris to those who believe that nature can be fully ensnared."[45] There is a reality beneath or under our cultural perceptions about and experiences in nonhuman nature. Nature is not infinitely malleable without risking destruction and loss. Nature, including wilderness, is a real place beyond human mental constructs and "our ability to cocreate."[46] Ecosystem entities with their integrity and history remain significant for ecological and perhaps cultural reasons, even in today's highly managed and technological world.

"Restoration practice is always about assisted recovery and not about the creation of artifact even if the deliberate (or otherwise) remnants of human activity are later evident."[47] Without natural processes, restoration is insignificant, inauthentic. What restorationists need to be able to do is to examine critically the "extent and thickness of the cultural layers we impose on top of reality."[48]

Restorationists attempt to observe and study ecosystems as objective entities, with their own evolutionary history and capacity for self-organization. They do so in order to model and copy, as best they can, the type of natural landscape under consideration. Hence, restoration for many is fundamentally a noncreative act in its objectives; it neither attempts to improve nor improvise on nature but only, as William Jordan states, "blankly, to copy it."[49] In this way the aim of restorationist, ecologist, practitioner, and volunteer alike is to let nature work through them but on nature's own terms. Through attentive, patient observation of and participation in nature's ways, restorationists attempt to allow land to speak in its own language.[50]

Yet restoration at its best also involves an attempt to reflect intentionally and critically on appropriate action in relation to particular landscapes. Insofar as it promotes this type of critical reflection, restoration practice can serve as a purifying tonic for stable, strategically essentialist, preformed assumptions about nature—and about the sacred. For working actively with often-diverse groups of people to restore land's health tends to be a destabilizing experience full of unexpected surprises that can generate new forms of awareness in relation to nature, people, and the sacred dimensions of embodied life. The ways in which restorationists interpret their experiences are, of course, shaped significantly by the intellectual and emotional frameworks they bring to such encounters. Still, it is also often the case that our previously held beliefs and assumptions about land, people, and the divine may, and often do, change or "bloom" in and through ongoing material encounters with the natural world and with diverse groups of people.

The restorationist may come to the field, for example, with certain assumptions about land's capacity to renew itself over time that are then changed when she is experientially drawn into nature's own slow, healing ways. She may be surprised by the extent of nature's gracious abundance and provision or, conversely, its callous indifference to suffering. Additionally, she may be inspired by the capacity of people to work together to jointly achieve a task; alternatively, she may be disheartened by how difficult and tiresome it is at times to work in community with others.

Do restored ecosystems finally represent cultural artifacts, creations of human tastes and desires, or natural entities, recoveries of land's intrinsic processes and functions? On the one hand restoration can never entirely escape its character as a human-initiated and, thus, cultural activity. An old-growth forest is not the same, or as valuable ecologically, as a hundred-year-old replanted one. Yet, on the other hand, restoration can and should attempt to attentively replicate natural systems as they once functioned so that nature is allowed to remake itself and regain its evolutionary path and future. An ecosystem in the process of being restored may begin in a sense as a cultural project, but in the end, or somewhere in the middle of the process, land becomes once again a dynamic, energized biotic community, taking off in its own direction.

The restorationist's experience between the worlds of constructivism and essentialism can serve as a fertile basis for the development of a complex meaning for nature. Critical of preexisting assumptions and open to evolving formulations, the restorationist may enter public environmental discourse ready to make a strong case for the inherent, perhaps even sacred, value of nature and the imperative for its healing. Several of the dimensions of a restoration formed meaning for nature deserve articulation. We turn next to this task.

DIMENSIONS OF A RESTORATION MEANING FOR NATURE

At least six dimensions in the complex meaning for nature "bloom brighter" in the light of restoration issues and activities. These are the elements of integrity, historicity, interrelatedness, self-healing capacity, wildness, and sacredness. In addressing them I consider both restoration and Christian environmental perspectives. As we go along I highlight some of the ways in which these concepts and ideas connect to and undergird a strategic essentialist model of nature.

Integrity

Ecological restoration has a preference for wholes—whole ecosystems and even larger natural landscapes that may involve multiple overlapping ecosystems. It attempts to put everything back—for instance, soils, plants, animals, that were once present so that an ecosystem can recover the natural process and functions it needs to return to its natural, historic trajectory prior to disturbance. The *sine qua non* of ecological restoration is recovery of natural processes, and whole intact eco-systems are necessary for these processes to commence. Following Aldo Leopold's famous dictum, most restorationists believe that "a thing is right when it tends to preserve the integrity, stability, and beauty of the biotic community. It is wrong whether it tends otherwise."[51]

The notion of ecological integrity is understood by many restorationists as "an all-encompassing term for the various features—resiliency, elasticity, stress response, and so on—that allow an ecosystem to adjust to environmental change."[52] In this way, ecosystems are defined according to certain objective, essentialist attri-butes that contribute to land health (or pathology). Additionally, it connotes an ecosystem's "original" integrity, that is, the features of the ecosystem's state prior to disruption. Echoing conservation biologists Paul Angermeier's and James Karr's classic definition of biological integrity, ecological integrity too implies that sys-tems are intact, with "native species populations in their historic variety and num-bers naturally interacting in naturally structured biotic communities."[53]

Ecological integrity for some restorationists also, however, involves a cultural dimension. This is particularly the case in contexts where cultural landscapes are the historical norm, or where human survival depends intimately on the restora-tion of damaged ecosystems. On this point, for example, the SER International Primer on Ecological Restoration states: "The North American focus on restoring pristine landscapes makes little or no sense in places like Europe where cultural landscapes are the norm, or in large parts of Africa, Asia and Latin America, where ecological restoration is untenable unless it manifestly bolsters the ecologi-cal base for human survival."[54] In this case, the notion of ecological integrity is importantly shaped by particular human needs and cultural perceptions, accentu-ating the constructivist aspect of a restoration meaning for nature.

Christian environmental thought, too, describes nature in terms of its inher-ent integrity or wholeness.[55] Here, however, the notion of the integrity of creation refers to earth's created wholeness and God's relationship to it. In this view it is God, the Spirit, or the divine that creates nature with an encompassing, organic integrity. For some, the integrity of creation is thought to originate in the very being of God prior to the birth of creation, giving it a fundamentally and essen-tially sacred meaning. Creation "possesses an inner cohesion and goodness" based on its origin in "the will and love of the Triune God," states Larry Rasmussen.[56] It is not humans who integrate creation; creation's integrity "is prior to our concern, prior to our participation."[57] Although humans are viewed as part of God's whole and intact earth, the focus for this understanding of creation's integrity is often on nonhuman nature's independent relationship to the creator. Accordingly, inso-far as humans defile or violate creation's inherent, God-given integrity, they act unfaithfully to God in this perspective. Emphasizing the strategic aspect of this view, these Christian environmentalists assert that ecological loss and degradation not only represent a sin against nature and its wild others, but they also "violate a divine relation, diminish the sacred, and offend against the Creator's love."[58] Where creation's given sanctity is damaged by human activities and still can be put back by human activities, it should be—as an act of obedience, service to or friendship with the God who created it in the first place. Where "original" ecological and biological integrity has been destroyed once and for all, for instance, when a whole species line is extirpated, or an entire ecosystem is wiped out and can never be recovered, it should be mourned and lamented as an act of unfaithfulness, disservice, or disloyalty to God, according to this particular Christian position.[59]

Christian ethicists who emphasize earth's limits and potential destruction by erring humans also tend to stress the belief that God, in some sense, continues to sustain the integrity of creation in and through God's ongoing activity in the world. So, for example, although James Gustafson's God is a powerful Other bearing down upon creation, God is also a powerful Other sustaining creation. God in this dual sense refers to "the powers that have brought life into being; that order the range of 'objects' and experiences," and "God's ongoing creative and ordering activity within the world, upon which we and the nonhuman natural world are dependent for the sustenance of life."[60] The question more precisely of how God sustains creation's integrity is an important question that theologians have thought long and hard about for centuries.[61] The central point of this first element in the meaning of nature, nevertheless, is that nature's moral standing is based on its evolutionary telos and/or divinely created integrity. Damaging nature represents a violation of land's intrinsic wholeness; restoration represents a strategic attempt to make nature once again whole.

Historicity

One reason restoration emphasizes ecological "wholes" is its emphasis on "original" or historically intact ecosystems prior to human disturbance. The reliance on history in ecological restoration, however, poses significant theoretical and practical challenges. For example, restorationists face the perennial difficulty of determining the reference conditions upon which the present restoration is modeled. How far back into the past should the restorationist go in gathering historical information about a particular natural landscape? In new world settings such as North America and Australia, restoring to an "original," "historic," or predisturbance condition tends to mean the state of an ecosystem prior to European settlement. Yet even this poses difficulties in terms of determining the extent and ways in which indigenous peoples inhabited and influenced ecosystems prior to colonization. And what if a natural area that was once inhabited by indigenous peoples is now a designated wilderness area—in restoring it, should humans be "reintroduced," along with, say, trees and wolves?

Determining the historical reference conditions upon which the restored eco-system is based is one of the most challenging aspects of restoration work. On the one hand the concept of reference conditions "is intuitive," as Higgs writes. It involves gaining "evidence from the past, as detailed as possible, that provides a singular portrait of the past as a goal for the future."[62] Nature has a distinct (essential) reality, a history and a trajectory that can be discovered and recovered, in a sense, through ecological restoration. The gathering of historical evidence takes many forms in ecological restoration practice: "baseline studies, control plots, interpolation and extrapolation of historical data, paleoecological studies, exclosure studies, and so on."[63] In indigenous restoration efforts, it also may involve gathering information from tribal elders or persons with traditional ecological knowledge (TEK) of plants and animals, or the contours of land and waterways, as they existed in the locale in the past.[64]

Despite these challenges, understanding nature in terms of its particular historical trajectory and character remains a central theme for most restorationists. Restoration, for instance, allows us, even in highly urbanized contexts, to attempt to maintain a degree of continuity with particular landed places, including the particular natural processes and species that once resided there. Different from the practices of, say, revegetation or reclamation, ecological restoration aims to regenerate historic or classic ecosystems, replete with their diverse natural features, in a way that reinstitutes land's naturally evolving trajectory.

Several years ago I attempted to research the natural history of the piece of land that surrounded our home in central Illinois. It was a farming area, and had been since European settlement of the area at the turn of the nineteenth century. The majority of "ground," as farmers referred to the land, was used for planting corn and soybeans, and, very occasionally, grasses such as alfalfa (for hay). Our home, a restored barn, sat on seven acres, a former planting field that had been converted to Kentucky bluegrass by the previous owners. Beyond our yard stretched thousands of acres of corn and soybean fields in every direction. Occasionally, a farmer had left an old hedge of Osage orange trees and prairie grasses to separate fields, though this was rare.

My interest in the restoration of the yard around our home was to provide a richer habitat for wildlife and insects, as well as a more aesthetically appealing natural area. Before deciding what to restore the lawn *to*, however, I wanted to learn what the land was like prior to European settlement and intensive farming of the area. I went to the library of natural history at the University of Illinois where all the state land surveys and natural historical data were housed. With the help of one of the reference librarians, I ended up finding a variety of interesting documents, including land plats back to the late 1800s. None of the documents, however, provided a coherent or conclusive picture in terms of specifying the exact ecosystem that had existed prior to the corn and soybean that preceded the bluegrass.

Typically, a restorationist will attempt to determine the natural and historical variability of an ecosystem over a set period of time (100, 1,000, or 10,000 years, say), and the ways in which cultural practices have (or have not) influenced the ecology under question. "The challenge for the restorationist lies in selecting an appropriate boundary around such variation, realizing that too narrow a choice, such as interannual changes, might overlook grander processes, and too large a time interval will displace fine-grained phenomena."[65] Approximation, writes Eric Higgs, "may be the best way to describe use of historical reference information." Reference information will require determining a range of variation in the historical dimensions that are utilized.

In the case of my restoration project, general natural historical information of the area told me that there were patches of woodland and oak savanna mixed among tall grass prairie. Getting the seven acres back to exactly what was there in the past probably was not as important as reflecting upon which ecosystem I would best be able to restore given technical and economic resources. I could simply let the sod grass grow and then wait to see what came back eventually; perhaps over time birds would carry in some plant seeds from the ridge way up behind the house or from our neighbor's farm (though it also had very few trees). I did this for two years. But the Kentucky bluegrass, not really meant for growing tall or allowing other species to coexist, became excessively thick and matted, prohibiting any other plant species to emerge. Nonetheless, perhaps over the course of fifty or a hundred years some trees would have grown, and it would have been an interesting experiment to see what the property would have turned into if I did finally just "let it go."

In the end I chose to restore prairie, if only because there was so little of it that had been regenerated in my farming community. Additionally, two of my farmer neighbors had complained that their apple trees had not been pollinated sufficiently the past two years, and I suspected the prairie, with its appeal to multiple bee species, could help with this problem. I was also motivated by the fact that I think that prairie, with its great variety of grasses and showy flowers, is just so lovely. These same neighbors offered to help with equipment (despite the fact that they thought I was crazy for wanting to put back "them damn weeds"); a locally based state conservation officer donated the herbicide to kill the bluegrass; the nonprofit organization Pheasants Forever gave me a variety of seeds in the interest of regenerating wildlife (pheasant) habitat (an interesting paradox, given that pheasant are not a native bird species to North America); and my children could help eventually with the prescribed burning that prairie grass requires to thrive.

The planning portion of the project took far longer (at least one year) than its implementation (three months), which is often the case in restoration work.

I spent a significant amount of time, for instance, researching whether it was necessary to use herbicide to kill the existing grass prior to planting the prairie grass seeds. I did not like the idea of using it both for health and philosophical reasons—it felt antithetical to the very idea of restoration to introduce a chemical that destroyed plant life and was potentially harmful to animal life. Even though the conservation officer explained to me that the type of herbicide (Roundup) that she recommended had a low runoff rate (the distance that a chemical "travels," through rain water and erosion, from where it is sprayed) and a relatively short half-life (the time it takes for the chemical to decay spontaneously), it nonetheless made me uneasy.

The herbicide, however, would kill the roots of the bluegrass, which was necessary for the prairie grasses to take root and flourish. Plowing up and turning over the grass would only activate the underlying seedbed, potentially causing other plants to grow and outcompete the prairie plants. If not herbicide, something would still need to be done to kill the existing grass without stimulating the growth of other plants. A fellow restorationist suggested the idea of laying down black plastic over the grass for several months in order to kill it. This tactic, however, seemed impractical at best and risky at worst given the size of the area and the chance that the plastic cover would not fully kill the roots. I did not want to half-kill the grass, only to have it come back and take over the newly planted prairie plants; that would be a waste of seed, money, and time. The herbicide, in the end, appeared to be the best choice in terms of ensuring that the bluegrass would be killed down to its roots, and that the prairie plants would successfully establish themselves.

Prior to applying the herbicide, the grass, which had grown tall and thick, needed to be cut close to the ground in order to allow the chemicals to penetrate the roots. It was early spring, the best time for planting prairie seeds, though this also posed challenges in terms of timing the mowing and then spraying. So that the grass would be as short as possible at the time of the spray, we would need to mow one day and spray within a couple days following. To spray successfully, it needed to be dry and still in order to minimize the chance that the herbicide would "travel," either through water or wind.

Optimizing these conditions proved difficult in central Illinois in mid-April. The first time that we mowed, the weather forecast was incorrect and it rained the entire week, causing the grass to grow fast and become too long to spray. We re-cut the grass and were forced to wait a couple of days due to high winds, though then finally proceeded with the first round of spraying. In order to completely kill the grass, two rounds of herbicide application, spaced approximately a week a part, were required. My farmer neighbors applied the Roundup using a small sprayer that they pulled behind a four-wheeler.

Approximately four weeks later, when the grass had yellowed sufficiently, we prepared for planting. Pheasants Forever loaned us a special planter with extra small holes in the drill (the part of the planter that bores the seed into the ground) so that the smallest prairie seeds would not drop down too quickly. My neighbors came over once again, this time with their smallest hydraulic tractor to pull the planter and drill in the seeds. I had several twenty-five-pound bags of mixed prairie seeds from Pheasants Forever that we emptied one at a time into the bin on the top of the planter. The farmers had never used a drill with such small holes—corn and soybeans are big seeds—and they were not convinced that the seed would actually drop down through the channels of the blade to be "cut into" the soil. So we tested the operation. They drove slowly for about fifteen feet while I stood on the back of the planter to make sure the seed was dropping down into the drill. Then they stopped and we all knelt down on hands and knees, opening the slice in the dirt with a jackknife to make sure that the seed had been released. Sure enough, there were the tiny prairie seeds, firmly in place in the soil.

Six weeks later, the prairie plants began to show themselves. By the end of the summer, we had our first peek at a fledgling prairie. I was not able to come close to putting back all the species—hundreds, close to a thousand, of them counting plants, insects, and animals in the historic tall grass prairie ecosystem—that once dominated the "original" midwest landscape. But I attempted, within constraints, to restore a piece of the historic landscape as adequately as possible.

Preferences for regenerating certain historic ecosystems are always to a certain extent contingent on cultural beliefs and experiences, as a strategic essentialist model of nature admits. We desire to restore wilderness areas in the United States, for example, because Americans have over time come to believe that wilderness is a good idea—perhaps "America's Best Idea," as documentary filmmaker Ken Burns suggests—for nonhumans and humans alike.[66] Nevertheless, our desire for healthy natural landscapes that have some continuity with the past also may run deeper than this. It may, for instance, reflect an inherent, essential human need to connect with natural life processes and landscapes that have proved beneficial for human physical and emotional well-being.[67]

Midwestern restorationists may decide to regenerate tall grass prairie to the region because people have come, once again, to perceive the prairie landscape as beautiful. They also may choose to restore prairie for ecological reasons; that is, because the prairie ecosystem will take root more easily (than, say, tropical rainforest), given the particular environmental conditions. Yet they may also restore prairie because it serves as a connection with the natural (and cultural), evolutionary historical trajectory of the particular place.

Culturally and historically contingent beliefs certainly play a role in determining what counts as "nature" (why prairie instead of planting field), and which features (why big blue stem instead of corn) should be restored to this or that part of nature (why my yard instead of my neighbor's) in the first place. Still, as I have already argued, restoration loses its significance if it is not based on the attempt to study and copy naturally evolving processes and functions. Understanding nature in terms of its history is valuable to restorationists because it helps ecologically to get an ecosystem's historical integrity more accurately right, and it signals a commitment—which often involves a moral choice—to the continuity of evolved natural processes and species in particular places.[68]

Interrelatedness

The next dimension in a restoration meaning for nature is its interrelatedness. Restoration authors frequently emphasize the idea that human and natural systems are intrinsically interrelated, and can be further interwoven in ways that promote the mutual benefit of people and land. Restoration provides a context, according to these authors, for better understanding the ways in which cultural and natural ecosystems can become more sustainably interconnected. Restoration concepts such as landscape coevolution (Higgs), biocultural restoration (Martinez), and reinhabitation (Mills), for example, well represent responses to this element of nature's meaning. According to these concepts, the interrelated character of people and land involves multiple dimensions, from the personal to the communal to the cultural.

Christian environmental authors have their own ways of articulating the importance of interrelatedness between humans and nonhumans: "being-with" is creation's "own way of being," according to this perspective.[69] "Existence is coexistence;" creation, including human beings, is inherently, essentially social; life necessarily involves mutuality.[70] Humans, as well as all animals, to various degrees, need others for their physical survival and emotional wellbeing. Life on earth, as created by God, may best be defined in terms of "ontology of communion," writes John Douglas Hall.[71] Maintaining and fostering communion, creation's first, fundamental, and essential value, is the goal of this view of nature.[72]

Feminist theologians and ethicists in particular have emphasized the relational dimension of creation's integrity, though in a way that has focused on the concept of the relational self, expanded to include nonhuman nature. Recognizing the potential dangers of a view of the self that is absorbed into or superseded by other nature, Sallie McFague proposes a mature sense of self that is capable of both embracing intimacy with and recognizing difference.[73] McFague views the self-other relation in terms of friendship whereby one can see the other objectively as differentiated subject. In her model, which emphasizes both the essential and constructed aspects of sociality, the self has its own world, subjectivity, life-project, or interests, without feeling the need to fuse with the other. The mature self, in other words, can recognize and affirm, with "loving objectivity," the otherness of the Other. "Loving the other as subject," writes McFague, "is the very reason for desiring the knowledge to be objective: to be what is best for this subject in *its* world."[74] A subject–subjects model, according to McFague, emphasizes the "'in itselfness' of each entity we know."

A subjects-to-subjects model, writes McFague, is a "maze model"—rather than a "landscape model" where subjects peer out and over, surveying the other as object—where subjects move about in the world, "being touched by and touching many others."[75] The self-other relation is a "dance of interaction," as psychoanalyst Jessica Benjamin puts it, or a "dynamic autonomy," as scientist Evelyn Fox Keller proposes.[76] It is multiple rather than dyadic, viewing the self's relation to the various others she encounters along a continuum. According to this view of human beings' relatedness to nature, the ambivalent, troublesome dimensions of the self-other relation are but one aspect of our inherently variegated relationship with the world of nature and its beings. Relationships, human/human, human/nonhuman, and nonhuman/nonhuman, are interactive, rich, and diverse in this perspective. Humans and the natural world may be intrinsically, essentially interrelated, though they are also "stitched together imperfectly," as historian of science Donna Haraway writes.[77]

Insofar as nature's own interrelatedness, that is, the ecological connections and interactions that operate among diverse biological beings and processes, is damaged by human activities, so too are opportunities for experiencing human fulfillment. In Christian environmental perspective, the loss of nature's interrelatedness, and the human connection to it, also involves an impoverishment in the human relationship to the divine. It may be that ecological community and the interrelated connections that inherently exist among beings may be regenerated, as restorationists argue. Nonetheless, for restorationists, secular and Christian, loss of both intact ecosystems and human connections with them represents a defilement and breakdown of the inherent capacity to relate to the wild others in our midst. From the perspective of strategic practice, restoration is viewed as the regeneration of ecologically necessary connections among organisms in the biotic community and culturally valuable connections between people and land.

Self-Healing Capacity

A dimension of our understanding of nature that restorationists emphasize repeatedly is the extent of land's capacity to heal itself if given the chance. In some instances, land will be so impoverished, stripped or blitzed from mining or nuclear testing or manufacturing, that there will be nothing left to regenerate. In these cases, humans may be able to revegetate or reclaim certain natural features, in effect creating a new natural landscape, though they will not be able to regenerate or restore biological processes that were once there.

Regeneration, on the other hand, implies that there is some semblance of resident natural value and dynamism remaining, even if only down deep in the soil. Here the reality of nature's inherent, essential proclivity toward self-renewal is accentuated. For in many instances, the energy of land's natural processes and historic trajectory has simply become blocked or otherwise rerouted by past human activities, and restoration can help get it back on track. The particular ecosystem's requisite parts—soils, waters, seeds, plants, animals—will need to be regenerated in their appropriate numbers and frequencies—enough for species' lines and genetic information to get up and going again, capable eventually of recovery and self-renewal.

In a coastal area of northern California, for example, old-growth redwoods, whole forests of them, were removed through intensive logging in the region over the past century. As a result, humidity and collection of moisture from sea fog was dramatically reduced, especially over the past fifty years, and desertification has begun to set in—a pattern familiar in many Western mountain regions. This reduction in the moisture-trapping canopy of the giant Redwoods meant that sea fog could no longer be trapped and then dropped to the ground in order to form springs. The loss of water created a cascade of other ecological losses: "Martens and fishes, salamanders, insectivorious birds and the notorious spotted owl—are missing or present only in much reduced relic populations. Salmon and steelhead are nearly gone."[78]

Restoration of the water was the main goal of (7,100 acre) restoration efforts in the region (facilitated by the already cited Intertribal Sinkyone Wilderness Council). "Everything else will follow from that," writes restoration ecologist Dennis Martinez. Restoring the water, in this case, involved the slow and laborious process of thinning sprouts and replanting seedlings. Existing redwood stump sprouts needed to be thinned (from seven to fifteen sprouts to one or two per stump) in logging areas where they had come up on their own. Nursery-grown redwood seedlings were hand-planted where sprouts had failed to self-regenerate due to intensive slash fires or the compacting weight of heavy logging equipment.[79]

Reduce and reform the logging practices, thin sprouts, replant seedlings, reintroduce some native grasses in order to encourage the emergence of larger forbs, burn selectively (perhaps not for 100 years in the Sinkyone case), let the sun shine, rain fall, and over time, the land begins to heal. Insects, birds and animals, some new and some old, find a way back; wildness asserts itself and land begins to regenerate. Humans choose consciously to engage in restorative practices, step in, for instance, to "return some otters, locally extinct, and put them back in the rivers," as philosopher Holmes Rolston cites.[80] Though "after a few generations," nevertheless, "the otters do not know they were once reintroduced; they behave instinctively, as they are genetically programmed to do. They catch muskrats as they can; population dynamics are restored and natural selection takes over."[81] In other words, humans decide morally and strategically to engage in ecological activities, but then they step back, and let nature regain its own, intrinsically self-renewing course.

According to this view, restorationists "facilitate nature, help it along, mostly by undoing the damage that humans have introduced, and then letting nature do for itself."[82] They are really more like midwives than artists or engineers. It's like the doctor used to say in the old days when setting a broken arm or leg: "Really, I just treated you, and nature healed you."[83] After nature does its healing, we do not consider the limb fake or artificial. And we do not view the doctor as having created an entirely new arm. The arm or leg healed itself, and the doctor facilitated it. With some assistance, natural processes can return to land, and land can heal itself.

Wildness

A variant of land's inherent capacity to heal itself is the restoration idea that nature is intrinsically "wild." Wildness connotes a quality of evolutionary biological life that is ultimately beyond human control and domestication. In defining this element of nature's meaning, restorationists often distinguish between "wildness" and "wilderness." This is because the notion of wilderness (e.g., Yellowstone National Park) implies that a natural area is cordoned off from human inhabitation and development, whereas wildness (my restored prairie) suggests that people can inhabit ecosystems, albeit in ways that promote "modest, regenerative, respectful activities over long intervals in precious areas," while still maintaining land's autonomous dynamism.[84] "The question," states Eric Higgs, "should not be, 'can we restore wilderness,' but can wildness be restored?"[85] It is the recovery of naturally evolving, self-renewing, thriving, wild natural lands—in wilderness, urban, rural, and/or suburban areas—that is the goal of restoration according to this element in the meaning of nature.

Restorationists who ascribe to a deep ecological philosophy (termed "Deep Ecological-Bioregional" restorationists in my field guide in chapter 1) in particular emphasize nature's inherent wildness as a distinct "reality" deeper than human perception. They characterize wildness in terms of nature's intrinsic evolutionary capacity for self-renewal and development, even realization in a sense. Humans too, as fundamentally wild animals, are characterized by intrinsic wildness. "There is a quality of wildness," writes Mills, "in every living organism, in every great planetary process—weather, tectonics, cell division, instinct."[86] In this way, the dimension of wildness also reflects our experience of nature as continuous with the human mind. For deep restorationists, restoration represents a way for humans to attempt to reinhabit and, in a sense, "tap into" wildness, within their own selves, as well as within ecosystems.

Even as restorationists lament losses in nature's intrinsic wildness, they are nevertheless often eloquent narrators, as we have already seen, of the ways in which wildness "comes back" when species and ecosystems are regenerated. Many restorationists, for example, soberly celebrate and trust the seeds of wild life, and the potential for self-healing within land that remains, even where species have been lost. Nature is just so much bigger and Other than human beings that its wild processes are ultimately beyond human control and perception, according to this view. Even when land has become degraded, energy, nutrients, and genetic information remain in the system. Seeds of wildness remain; wild, prolific life may still reemerge. As long as there is biological life on earth, wildness cannot be completely destroyed. However latent, it remains a part of the meaning of nature.

Intrinsic-Sacred Value

Closely related to nature's wildness is the idea that nature in some sense is intrinsically valuable. Its meaning transcends any purely instrumental value for humans. It is in some sense sacred in itself. Restorationists have varying views, as we have seen, of what actually happens to nature's intrinsic value when natural values are lost or subordinated to human acts of degradation. Some restoration authors such as Rolston identify an objective loss in nature's true, original value where ecosystems are humanly damaged, while others emphasize the idea that nature's intrinsic, sacred, wild value can never be completely lost through human acts of degradation. Still others, such as William Jordan, posit that nature's sacred value is actually created by human beings when they consecrate ecosystems through ritualized actions such as restoration (see chapter 5). Each of these views, nonetheless, is importantly characterized by the idea that nature holds an intrinsic or sacred value—whether in an essential or a constructed sense—that should be respected as such.

In Christian thought, the notion that creation is in some sense infused with divine, sacred presence is often referred to as a sacramental or panentheistic perspective.[87] Here God is understood to reside within, underneath, above, and around earth and its beings. Although the divine itself is not identified with the creaturely in this view, the finite creation becomes sacred, holding the divine presence. Now "everything that has breath" praises God, as theologian Jurgen Moltmann writes.[88] The mountains sing God's praise, the trees clap their hands. Nature's "true" value is defined not only in ecological terms, but in spiritual terms; creation's integrity is characterized in terms of its capacity for internal self-renewal, but also in terms of the ways in which it harbors intimations of the infinite. This is the "ubiquitous God *par excellence*," writes Sallie McFague. "This God is never absent . . . we can experience God anywhere and everywhere."[89] Insofar as ecosystems and natural value are lost, so too are modes of divine presence, according to this view.[90]

Some ecological theologians have pushed deeper into Christianity's sacramental tradition, positing a divine, sacred status for nature and its beings in themselves. Mark Wallace, for example, draws on quasi-animistic or pantheistic dimensions of the Christian tradition in which the universe and its beings are viewed as a manifestation of the divine nature (the Greek term for pantheism translates "all divine"). Along these lines, Wallace points to the ways in which the Holy Spirit is symbolized biblically as earth, air, water, and fire, in other words, as the basic elements and processes of the physical world.[91]

To name the Spirit as "life-form," writes Wallace "is to signal the Spirit's identity as a living being, a being whose nature is the same as all other participants in the biotic and abiotic environments that make up our planet home. Running rivers, prairie fires, coral reefs, schools of blue whales, equatorial forests—the Spirit both shares the same nature of other life-forms and is the animating force that enlivens all members of the lifeweb."[92] Along similar lines, ecofeminist theologian Ivone Gebara argues that God is "that sacred energy which pervades all beings."[93] Human persons know this relatedness, this sacred energy, in a personal way and thus it is appropriate to think of God, in an analogous way, as a person (though God is also more than personal). "By analogy, God is a human person, the sap of human life, but also the sap of the life in trees, in flowers, in animals, and in all that exists. By analogy too, God is man, woman, breeze, hurricane, tenderness, jealousy, compassion, mercy: Mystery."[94]

Panentheistic and pantheistic perspectives differ in terms of the extent to which God and the natural world are viewed as connected realities. Panentheism, for example, views the natural world as both part of and distinct from the being of God, whereas pantheism draws no such ontological (categorical) distinction between the realities of God and nature. In panentheism, God is understood as both immanent and present within and transcendent and distinct from creation; in pantheism, God is viewed only as immanent within creation, nothing is separable from sacred reality.

Despite such differences, both panentheistic and pantheistic versions of Christian sacramentalism emphasize the idea that the loss of nature's value is a real loss of divine presence in the world. Additionally, both types of deep theologians suggest that the sacred energy pulsing through the earth and its processes and beings can and will never be wiped out as long as biological life exists on earth. Life on earth, according to this view, has a built-in propensity for novel, prolific, persistent forms of regeneration. Natural processes and, by analogy, the divine presence, in some sense, cannot be stopped. Although serving neither as a justification for destruction or apathy toward creation—"Who cares, there's nothing we can do anyway!"—this aspect of the view that the sacred infuses the natural world heightens the intensity of moral response. For now, nature's sacred dimension not only permeates each and every being, process, and function of earth, but it does so superpowerfully, beyond human reproach or control. Hyper-attentive to the divine presence within earth and its forms, sacred loss, and persistence, is both acutely noted and felt. Emphasizing restoration as a strategic, moral practice, it is viewed as a faithful response to the sacredness of the natural world.

CONCLUSION

The notion of "nature" constitutes a central even though contested concept for restorationists. Key dimensions in the meaning of nature for ecological restoration are its integrity, historicity, interrelatedness, self-healing capacity, wildness, and sacredness. These elements of meaning are gleaned and shaped through the concrete processes of working to heal damaged natural lands. They are rooted solidly in the characteristics of land itself, as well as in ongoing adaptive stages that shift with the dynamism of nature. I move on in the next chapter to consider the personal and communal experiences of transformation and renewal that restoration practice may yield. It may be that we will want to restore because nature is understood as having an inherent integrity and wildness that is worthy of regeneration. Yet it may also be that we will want to restore for experiential and emotional reasons; that is, we may want to restore damaged land because our spirituality, our humanity, depends on a thriving relationship with a healthy natural world.

NOTES

1. For essays by Naess, Stone, Singer, Callicott, and Rolston, see the anthology edited by Zimmerman, Callicott, and Warren, *Environmental Philosophy*. On the bioregional view of Berg, see his edited volume, *Reinhabiting a Separate Country*.
2. Elliot, "Faking Nature," 81.
3. Ibid.
4. Ibid., 82.
5. Ibid., 81.
6. Ibid., 83.
7. Katz, "The Big Lie," 234.
8. Ibid., 232.
9. Ibid., 234.
10. Katz does not want to simply follow Elliot's use of the term "artifact" to refer to restored nature because, on the flip side, he does not believe that *undisturbed* nature can be compared to a work of art. "Natural entities have to be evaluated on their own terms, not as artworks, machines, factories, or any other human-created artifact." Ibid., 234–35.

11. Ibid., 239.
12. Ibid., 240.
13. Ibid., 232.
14. Ibid., 235.
15. Ibid.
16. Ibid.
17. See Mill, "Nature," 372–402.
18. Katz, "The Big Lie," 239.
19. Ibid., 240.
20. Ibid.
21. See, for example, Callicott, "La Nature est morte," 16–23.
22. A prime example of such counterproposals to postmodern deconstructionist views of nature can be seen in the collection of essays in Soule and Lease, *Reinventing Nature?*
23. McQuillan, "Defending the Ethics," 27–31.
24. Ibid., 29.
25. Ibid., 31.
26. Ibid.
27. Ibid.
28. Ibid.
29. On environmentalists who have debunked postmodern deconstruction thought on grounds McQuillan suggests, see Soule and Lease, *Reinventing Nature?*
30. Restorationists in the essentialist/constructivist debate tend to use the notion of the real—as that which exists "beyond" or "other than" the human mind—in two related ways: First, it can refer to the idea that there is an objective reality, namely natural processes and functions, that is not dependent on epistemology and social construction by humans; second, it can signal the notion that natural processes and functions are ultimately beyond the capacity for humans to control, manipulate, or destroy them.
31. McQuillan, "Defending the Ethics," 31.
32. Davis and Slobodkin, "Science and Values," 1.
33. Ibid.
34. Ibid.
35. Ibid., 5.
36. Ibid.
37. Ibid.
38. On this point regarding Christian ecological theology and deconstruction, see Wallace, "Sacred-Land Theology," 291–314.
39. Wallace, "Sacred-Land Theology," 306.
40. Ibid.
41. Callicott, "La Nature est morte," 16–17.
42. Higgs utilizes the term "critical constructivism" to refer to restoration understandings of "nature," though I prefer the notion "strategic essentialism" given its emphasis on nature's intrinsic values and capacities for self-renewal. Theologian

Serene Jones utilizes the term in relation to women's inherent capacities in a way similar to my own. See her *Feminist Theory and Theology*.

43. See Winterhalder, Clewell, and Aronson, "Values and Science," 4.
44. Ibid.
45. Higgs, *Nature by Design*, 179.
46. Higgs quotes Wordsworth's "Tintern Abbey" on this point. Ibid.
47. Ibid., 202.
48. Ibid.
49. Jordan, *The Sunflower Forest*, 76.
50. Ibid.
51. Leopold, http://www.us.oup.com/us/catalog/general/subject/LifeSciences/Ecology/ConservationBiology/?view=usa&ci=0195053052. *Sand County Almanac*, 262.
52. Higgs, *Nature by Design*, 122.
53. Angermeier and Karr, "Biological Integrity," quoted in Higgs, *Nature by Design*, 122. This focus on original, intact ecosystems, replete with indigenous species populations in their historical variety and numbers, gives ecological integrity an edge over ecological health, in my view, at least as a sort of lodestar concept.
54. Science and Policy Working Group, "The SER International Primer," 2.
55. See, for example, Rasmussen's *Earth Community*, McFague's *Life Abundant*, and Bouma-Prediger's *Beauty of the Earth*.
56. *Integrity of Creation—An Ecumenical Discussion*, Granvollen, Norway, February 25–March 3, 1998 (Geneva: World Council of Churches/Church and Society, 1989), quoted in Rasmussen, *Earth Community*, 105.
57. Ibid.
58. Ibid., 64.
59. Ibid., 101.
60. Gustafson, *Ethics from a Theocentric Perspective*, 239. See also Gustafson's more recent book on the environment, *A Sense of the Divine*.
61. For example, Gustafson is adamant that God's ongoing sustaining activity in the world does not suggest a divine telos for creation, and "most certainly not a *telos* that guarantees that the prerequisites exist for the sake of our species or any one of us as individuals." Gustafson, *Ethics from a Theocentric Perspective*, 240. Other Christian theological ethicists would disagree with Gustafson's position on this particular point. Still, his overall project to construct a less anthropocentric Christian theological ethic resonates with many Christian environmental authors.
62. Higgs, *Nature by Design*, 158.
63. Ibid.
64. See, for example, the Sinkyone Intertribal restoration project highlighted in Martinez, "Northwestern Coastal Forests," 64–69.
65. Higgs, *Nature by Design*, 164.
66. Duncan, "National Parks."
67. On this notion, see Kellert and Wilson's work on the biophilia hypothesis, for example, their *Biophilia Hypothesis*. See also Kellert, *Birthright*.

68. As restorationists O'Neill and Holland write in relation to an ancient woodland, we value it because "it embodies the work of human generations and the chance colonization of species and has value because of the processes that have made it what it is. No reproduction could have the same value, because its history is wrong." O'Neill and Holland, "Two Approaches," in *Cultural and Spiritual Values* and quoted in Higgs, *Nature by Design*, 154.

69. Rasmussen, *Earth Community*, 106.

70. Ibid.

71. Hall, *Professing the Faith*, 317, quoted in Rasmussen, *Earth Community*, 106.

72. Rasmussen, *Earth Community*, 237.

73. See McFague, *Super Natural Christians*, 103–7.

74. Ibid., 107.

75. Ibid., 105.

76. Benjamin, *The Bonds of Love*, 53; Keller, *Reflections on Gender and Science*, both quoted in McFague, *Super Natural Christians*, 104.

77. Donna Haraway, "Actors Are Cyborg" in *Technoculture*, ed. Constance Penley and Andrew Ross (Minneapolis: University of Minnesota Press, 1991), 22.

78. Martinez, "Northwestern Coastal Forests," 67.

79. Ibid., 68.

80. Rolston only minimally refers to the issue of non/native species in restoration efforts. Still, based on his concern for the conservation of naturally historic values, resident within particular places (e.g., rare mosses indigenous to the Rocky Mountains) we can speculate that where possible, Rolston would prefer the reintroduction of indigenous otters. Rolston, "Restoration," 129.

81. Ibid.

82. Ibid.

83. Ibid.

84. Ibid.

85. Higgs, *Nature by Design*, 58.

86. Mills, *In Service of the Wild*, 20.

87. For more on the idea of sacramentalism, see the section on "Defining Nature-Based Spirituality" in the following chapter.

88. Moltmann, *God in Creation*, 71.

89. McFague, *Life Abundant*, 149.

90. As Thomas Berry states, "We should be clear about what happens when we destroy the living forms of this planet. The first consequence is that we destroy modes of divine presence." See his *The Dream of the Earth*, 11.

91. See Wallace, *Finding God in the Singing River*.

92. Ibid., 38.

93. Ivone Gebara, *Longing for Running Water*.

94. Ibid.

RESTORATION OF THE PERSONAL HEART

Toward a Spirituality of Environmental Action

I believe quite sincerely that in these difficult times we need more than ever to keep alive those arts from which [we] derive inspiration and courage and consolation—in a word, strength of spirit.

—Rachel Carson, *Silent Spring*

It is a warm, late May day in Vermont and I am sitting around the kitchen table of Marty Illick, director of the Lewis Creek Association (LCA), eating a lunch of squash soup, tomatoes, and apples, all from Illick's backyard garden. The LCA is a community-based organization that formed two decades ago in order to organize and educate residents regarding the restoration, protection, and care of the Lewis Creek watershed in central Vermont. Given the LCA's reputation in Vermont conservation circles for working on watershed restoration from a holistic, cultural, and ecological perspective, I was meeting with Illick, along with two other women integral to the LCA operation.

Stream bank restoration, sustainable forestry practices, water monitoring, land and wetlands conservation methods, stream corridor analyses, and wildlife tracking are among the restoration-oriented activities in which the LCA is involved. The particular weekend that I was visiting, Illick and others from the LCA were involved with a restoration project that involved removing an invasive plant, European frogbit (*Hydrocharis morsusranae*), from local lakes and wetlands. European frogbit, a nonnative aquatic plant, causes native vegetation populations and diversity in wetland habitats to decrease and is also suspected of diminishing oxygen levels, which directly affects habitat for spawning fish and benthic organisms.[1] Frogbit was first observed and documented in the area in 2007 at a Natural

Heritage site in Town Farm Bay along Lake Champlain. By 2009 frogbit was estimated to cover 50 percent of the Champlain bay area.

Removing frogbit is tedious. Crews of conservation professionals and volunteers work out of canoes and kayaks, as well as walk in the shallower water with chest or hip waders. The preferred method of removal is hand harvesting with a metal gardening tool and a small, long-handled bamboo rake to reach both individual and larger mats of plants. Kayaks are fitted with plastic laundry baskets affixed with bungee cords to the bow of the boat in order to contain harvested frogbit; canoes carry five-gallon plastic buckets with holes drilled in the bottom to allow water to drain.

After plants are hand harvested into the boats' buckets and baskets, they are then unloaded onto a scow, or floating dock, anchored in the wetland. The scows were donated by a nearby marina and adapted with walls and fabric to hold a week's amount of hand-harvested frogbit. Marina employees donate their time to tow and empty the filled scows when they become full. Frogbit plants are stored on upslope land near the marina, where it quickly decomposes. The composted frogbit is donated to local farmers and used as mulch and compost for their fields and gardens.

The LCA also has social and cultural dimensions and goals that are integral to its restoration work. Its mission statement reads, "The mission of Lewis Creek Association is to protect, maintain and restore ecological health while promoting social values that support sustainable community development in the Lewis Creek watershed region and Vermont."[2] Additionally, the LCA states that it draws inspiration from the principles of the global, religious-oriented "Earth Charter": Respect and care for the community of life, ecological integrity, social and economic justice, and democracy, nonviolence, and peace. I asked Illick what she thinks it is about restoration work in particular that gives it the capacity for fostering such deep social, ethical, and spiritual values in relation to particular landed places. She replied, "I'm not entirely sure. But it is the most uplifting thing. We are really just high being out there in nature, working with a small group of people."[3]

Restorationists such as Illick often describe subjective experiences of transformation and renewal that are formed through the process of working to regenerate damaged ecosystesms. Yet environmental writers, secular and religious, have overlooked these restoration-based experiences, either as a way to analyze emerging forms of nature-based spirituality or as a way to examine the implications of such profound and meaningful experiences for building a broader environmental culture. This chapter attempts to remedy this oversight by exploring the spiritual landscape of restoration experience, describing some of its key dimensions and their significance for an environmental ethic.

Building on the definitions of religion and spirituality given in the introduction, I begin by further clarifying what is meant by a notion of nature-based spirituality in particular. Since I am especially interested in examining the ways in which restoration spiritual experience and moral action are interrelated, I explore the potential connections among spirituality, nature, and ethics. Next, I propose several criteria that may constitute restoration as a type of spirituality of environmental ethical action. Finally, I describe some of the direct experiences of transformation and renewal that tend to characterize restoration spirituality, noting the ways in which these relate specifically to Christian understandings of nature-based spirituality. Based on the background this chapter and previous chapters provide, I propose in conclusion several precepts for a spiritual ecological view of land partnership. These precepts provide an initial framework for the restoration of symbolic and ethical values that are further proposed in chapters 4, 5, and 6. For now we turn to defining more explicitly the concept of nature-based spirituality.

DEFINING NATURE-BASED SPIRITUALITY

The natural world, as I suggested in the previous chapter, is an orienting reality for restorationists. This is also true for restoration-inspired spiritual experience. As is the case with general notions of spirituality (see introduction), understandings of nature–spirituality are pluralistic and multifaceted. For example, scholars variously refer to nature–spirituality as spiritual experience that grows out of certain outdoor activities (Samuel Snyder's spirituality of fly-fishing), radical environmental activism (Bron Taylor's dark green religion), the environmental movement (Roger Gottlieb's environmentalism as spirituality), human biology (Stephen Kellert's biophilic spirituality), or traditional religious life (Sarah McFarland Taylor's spiritual ecology of green nuns). While scholars such as these may use various terms to describe nature-based spiritual experience—for example, nature-as-sacred religion, religious ecology, moral ecology, and spiritual ecology—they nonetheless share the conviction that deep, meaningful, and sacred experiences of healing and renewal are born from direct encounters with the natural world.

Christian thought has developed various understandings and dimensions of nature-based spiritual experience. In the fifth century CE, St. Augustine, for example, emphasized the ways in which nature's beauty pointed to the divine. In the thirteenth century Thomas Aquinas wrote about the ways in which every being, human and nonhuman, participated in a God who is all being. Each being, according to Aquinas, has in some way instrumental value in relation to others and to the whole, yet every being also has beauty and goodness in itself by virtue of its participation in God.[4] And in the sixteenth century John Calvin understood creation as the medium through which we see "sparks of God's glory." Nature is God's theater and God's robe, according to Calvin; in it humans could come to experience and know divine presence in the world.[5]

In the twentieth century ecological theologians expanded upon traditional spiritualities of nature in order to consider contemporary scientific perspectives on the natural world and on human beings. Often emphasizing the theological idea of sacramentalism—the concept, as noted in the previous chapter, that the divine is present and discernible in and through the material, physical world— many of these authors also draw on an evolutionary cosmology to interpret Christian ecological spirituality.[6] In Christian thought sacramentalism traditionally refers to the idea that divine presence is revealed in the ritual practices and material elements (e.g., water, bread, wine) of the Christian sacraments (e.g., baptism and communion). Yet the sacramental trajectory of Christianity also historically includes the idea that "God exists in some meaningful way in all wheat and grapes, all hydrogen and oxygen, and indeed throughout the world."[7]

Further, although nature itself is not equated as divine in a sacramental perspective, it does suggest that "the infinite is a dimension of the finite; the transcendent is immanent; the sacred is the ordinary in another, numinous light."[8] Kevin O'Brien draws on this interpretation of sacramentalism, for instance, in explaining the importance and significance of biodiversity from a Christian perspective. "Sacramentality," he writes, "tells us that the creatures with whom we share the world and the diverse systems they inhabit are a sign of a connection to God."[9]

Ecological theologians and ethicists that draw on a sacramental spiritual view of the world suggest that it can inspire a sense of "wonder, awe, amazement, fascination, astonishment, curiosity, and surprise . . . a sense of being very small amidst a grand Reality."[10] Along these lines Sallie McFague, for instance, proposes a Christian horizontal sacramentalism where the Christian loving eye can become educated to see the "embodied, concrete particularity of all things" in God.[11] In this view, "all the bodies, from subatomic to galactic ones, and all the ones in between, from robins to tigers to mountains and oceans, from mites to microbes to trees and plants, as well as, of course, ourselves"—all of these "millions, billions of creatures and entities, are, like us, made for the glory of God."[12] Similarly, Larry Rasmussen calls for a sacramental spirituality, stating his preference for an evolutionary sacramentalist cosmology, in which the divine is recognized and celebrated as "in, with, and under all nature, ourselves included."[13]

Despite the plurality of ways in which nature spirituality has been understood within both secular and Christian environmental thought, common themes emerge. In the first place nonhuman nature itself is commonly understood as the primary ground, Other, or mediator of nature-oriented spiritual experience. In this view, although access to the sacred may be possible through various media, it is especially in encountering the natural world and its entities that people can experience clarity of insight, a sense of wholeness, the peace of God, the ground of Wisdom, or however spiritual experience is characterized in nature-based spiritual accounts. Black Elk's vision, for example, comes in the clouds and sky; Thomas Merton hears speech pouring down in the night rain; Annie Dillard sees clearly and luminously in locking eyes with a weasel; Hildegard of Bingen knows God aflame in the beauty of the meadows, gleaming in the waters, and burning in the sun, moon, and stars, awakening everything to life.[14]

Second, and closely related to the above, nature itself is thought to hold knowledge, wisdom, even truths, regarding the way humans can understand themselves. For example, nature-based spiritual experience can remind individuals of "truths that industrial civilization and many forms of patriarchy have obscured . . . that we are physical beings, made of the same stuff as earth and stream and air, or that we need wilderness because, as Edward Abbey observed, we ourselves are wild animals."[15] Further, nature-based spiritual experience is understood to enable people to move beyond conventional social understandings of the self or social constructions of the body and thereby to see themselves as part of, and importantly defined by, the larger ecological community of life. This dimension of spiritual experience often involves a de-centering and then re-centering of the self, whereby one experiences a sense of loss and then enlargement of the self in relation to a larger, though interconnected and interdependent, community of life on earth.

Finally, accounts of nature-oriented spiritual experience, as already noted with regard to nature-based notions of religion, tend to include a moral orientation. That is, the sense of kinship and dependence people come to feel and know in relation to the larger community of life may motivate, in turn, an attitude of care and respect in relation to the natural world. For instance, early environmentalist John Muir frequently narrated the connections between a deep spiritual attachment to the natural world and its beings and the respect and love that inevitably followed. Aldo Leopold narrated the ways in which love and affection for land were precursors to respect and active care. And pioneering ecologist Rachel Carson wrote about the relation between developing a sense of wonder in relation to the natural world in childhood and the ways in which this, over time, developed into a moral ecological sensibility.

Despite the positive accounts just cited, however, several questions remain: How can we be sure that the search for communion with the natural world and its beings, and its care, will go hand in hand? In other words, how can we be sure that nature-based spiritual experience will in fact lead to environmental ethical action, rather than becoming merely a preoccupation with gathering nature-based therapeutic experiences or an escape from public engagement or social life? More specifically, are there certain types or dimensions of spirituality that are better or worse for creating just, restorative, and meaningful relationships between people and land?

NATURE, SPIRITUALITY, ETHICS

Philosophers, theologians, and mystics alike have thought long and hard about the relation between the contemplative and active, spiritual, and moral aspects of human life. Aristotle, for example, argued that the contemplative life of seeking excellence and virtue (*arête*), and leisure, should be balanced by a life of action and function (*ergon*), and further, that virtue is found both in contemplative and active practices.[16] Augustine wondered about the nature of the relation between love of God and love of neighbor.[17] Dorothy Day, cofounder of the mid-twentieth-century Catholic Worker Movement, wrote about the ways in which living among and advocating for the homeless poor led one closer to union with God.[18]

Extending the thought of authors such as these, I would like to explore how an environmental spirituality may be linked to an environmental ethic. For while recent environmental writers, again secular and religious, have suggested that there may be a positive connection between nature-based spiritual experience and moral action, the more precise nature of this connection has been overlooked.[19] It may be that certain experiences of nature have the capacity to school a person in the virtues of humility, sufficiency, and attentiveness, for example. But more can be said explicitly about the ways in which nature-based spirituality and ethics are connected. There are at least three ways in which the connection between nature-based spiritual experience and the moral ecological life can be articulated.

First, and most basically, nature-based spiritual experience involves a sensory dimension that, in turn, can form a gateway for the formation of moral ecological action. Studies from various disciplines such as ecological psychology, evolutionary biology, child psychology, and social ecology increasingly show that the emotional basis for affection and care in relation to nature begins in childhood.[20] In this stage in human development one's sensory world is wide open and oriented to stimulation by the diversity and richness of the natural world. Moreover, ecological moral values are shaped integrally by the myriad sensory stimuli of the outer world.[21] "Show me the landscape in which you live and I will show you who you are," writes Spanish philosopher Jose Ortega y Gasset.[22] "Trace the history of a river, or a raindrop, as John Muir would have done," nature writer Gretel Ehrlich says, and you will also be tracing "the history of the soul, the history of the mind descending and arising in the body."[23] The physical earth's outer landscape and personal self's inner landscape, in other words, are integrally connected; who we become as human beings depends not only on what we think and do but where on earth our thoughts and actions are formed.

When children or adults encounter the divine presence in nature, as Black Elk, Merton, and Hildegard do, they encounter the sensual world of nature on their skin; they smell, taste, hear, and are taken over bodily by the presence that is around and upon them. "Cities give not the human senses room enough," writes Ralph Waldo Emerson.[24] It is the world of nature, as Emerson believed, that truly satisfies the depth of human sensory needs; nature, Emerson states, is "water to our thirst," "the rock, the ground, to our eyes, and hands, and feet. It is firm water: it is cold flame: what health, what affinity!"[25]

These sensory experiences and feelings in turn can instigate recognition, awareness, attentiveness, respect, and over time even love and care in relation to the natural world. In nature-based spiritual experience, one does not encounter the numinous dimension within nature abstractly; it is felt, smelled, and heard concretely in encountering other embodied creatures and biophysical processes in the natural world. The "music and pictures of the most ancient religion," states Emerson, for example, are heard and seen in "the fall of snowflakes in a still air ... the blowing of sleet over a wide sheet of water, and over plains, the waving rye-field ... the reflections of trees and flowers in glassy lakes; the musical steaming odorous south wind, which converts all trees to windharps; the crackling and spurting of hemlock in the flames; or of pine logs."[26] Concrete, tangible encounters with snow, sleet, wind, water, grass, trees, fire, all of these (and more), help to shape a person's attitudes and affections in relation to the natural world, according to Emerson.

The sensory encounter with the natural world may also elicit negative emotions and feelings. For instance, feelings of uncertainty, uneasiness, dread, discontent, and disconnectedness can afflict even the most ardent nature lover. Although mostly noted for his deeply romantic attitude toward nature ("never a day profane spent in nature"), Emerson also wrote about a restless discontent that inevitably invaded feelings of oneness with nature and its beings.[27] Hunters and anglers, for instance, often describe in detail a complex constellation of positive and negative sensory feelings and emotions that arise through the process of encountering and hunting another creature.[28] In relation to an elk hunt, nature writer David Petersen states that his "empathy is gut-churning": "As I gaze at this gorgeous wild beast I have so eagerly killed, my eyes cloud with tears, which I accept without shame. Yet at the same time I am positively electrified, buzzing with Ortega y Gasset's 'mystical agitation.'"[29]

Most ecofeminists critique hunting as a form of male aggression and violence against nature. Yet some have noted the deep feelings of kinship with the natural world and its beings, even the killed animal that can develop within the hunter. Ecofeminist Mary Zeiss Stange, for example, writes: "Far from being a mark of moral failure, this absence of guilt ... suggests a highly developed moral consciousness, in tune with the realities of the life-death-life process of the natural world. The simplistic analogy of hunting to such forms of male aggression as rape and warfare breaks down at precisely this point, where a kinship is perceived between hunter and hunted."[30] Negative feelings in relation to the natural world can be, in other words, intertwined with and lead to more positive ones. A sense of fear and awe in relation to the physical events (e.g., earthquakes, storms) within earth can form feelings of dependency and, in turn, the spiritual quality of gratitude before a larger, more powerful, and astounding nature.

This points to a second way in which nature-based spiritual experience can be understood to generate and shape moral ecological values. That is, nature-based spiritual experience has the tendency to form particular perceptions of earth and its beings. Feelings of deep connection can form, for example, a loving perception of the natural world whereby one comes to admire and respect nature's ways. Loving perception of nature also may be accompanied by the understanding that earth and its beings are in some sense sacred, worthy of being valued for their own sake as well as for their interconnected and interdependent character.

From this perspective the earth with all its life is taken, as Roger Gottlieb writes, "as an ultimate truth," a sacred, holy truth if you will.[31] "As more familiar mystical experiences might alter our attitudes toward death, our fear of the unknown, or our petty insecurities, realization of our kinship with the earth confirms the need to question any unquestioned trashing of what surrounds us."[32] Perceiving earth as sacred has the potential for motivating ethical action in response to nature, particularly when land and its beings are unjustifiably or unquestionably damaged or destroyed. In nature-based spiritual experience, the "earth is not just being polluted, it is being *desecrated*," Gottlieb writes. "Something more than useful, more than physically pleasing, something holy is being torn to bits for what are too often the more trivial, thoughtless, or downright cruel of reasons."[33]

A third and final connection to be noted between nature spirituality and moral ecological consciousness is a particular way of being in the world—formed through experiences of the surrounding world illuminated by natural light. As numerous nature writers attest, moral ecological virtue or character is often born through the course of nature's embrace. The world of nature brings out the best in human beings, to paraphrase Emerson.[34] The person who spends significant time contemplating the intricacies, rhythms, changes, beauties, sufferings, and patterns of a particular forest or woodlot or wetland comes to know that place and its beings from the inside out; it becomes part of her and she part of it. Lessons are learned, for example, about balance and integrity, the relation between suffering that is inevitable and suffering that does not have to be, patience and attentiveness. Human hubris and shortsightedness are juxtaposed with evolutionary historical time; the joys and travails of human life become intimations of earth's seasons.

But beyond the connections just outlined, nature spirituality may also involve a public ethical dimension or trajectory. Put another way, nature-based spiritual and moral experience, in addition to being importantly grounded in and shaped by communion with nature itself, may be equally grounded in and shaped by environmental ethical action.

A SPIRITUALITY OF ENVIRONMENTAL ACTION

Of course, not all spiritually motivated public action or spiritualities grounded in nature are equally welcomed. As history has shown, politics justified on spiritual or religious grounds can develop into a kind of fundamentalism and even dogmatic violence.[35] Spiritualities shaped by particular landed places and biological identities have at times historically morphed into faiths of "blood and soil" (as was the case with National Socialism in Nazi Germany). This sobers any romantic enthusiasm for nature-based spirituality and particular land-based practices that lack ethical norms or criteria. Further, this presses us to examine ecological restoration practices for their ethical soundness. In chapter 5 I propose a view of restoration as a type of ecological sacramental practice or a public spiritual practice. Here, as a precursor to this idea, I propose key criteria that constitute restoration as a type of spirituality of environmental action.[36]

At the outset, a brief note on what I mean by the phrase "a spirituality of environmental action" is required. A spirituality of environmental action is related to, yet extends, sociologist Robert Wuthnow's concept of practice-oriented spirituality. A practice-oriented spirituality, writes Wuthnow, involves engaging "intentionally in activities that deepen [one's] relationship to the sacred."[37] It also involves a communal, meaning-making function, according to Wuthnow. Some of the elements or particular practices involved in a practice-oriented spirituality, for Wuthnow, include prayer, devotional reading, and service of others; discernment; an orderly and regular approach to ritual; a revisioning of the self; a sharing of our experiences and stories with others; the use of a wide variety of resources for inner work; and an emphasis upon an ethical dimension. "The point of spiritual practice is not to elevate an isolated set of activities over the rest of life," writes Wuthnow, "but to electrify the spiritual impulse that animates all of life."[38]

Despite the biological resonance of the text just cited, Wuthnow considers neither the natural world nor environmental practices in his conceptualization of spirituality. Nonetheless, certain environmental activities and practices—as Bron Taylor and others write about in relation to, for example, surfing (Taylor), fly-fishing (Samuel Snyder), homesteading (Rebecca Kneale Gould), community gardening (Sarah McFarland Taylor)—clearly represent a practice-oriented spirituality. In fact, restoration practice may constitute a paradigmatic example of this, especially insofar as it meets specific criteria of a spirituality of environmental action. Implicit in my discussion of these is the assumption that the relationship between spiritual experience and ethical action is mutually generative: Spiritual experiences in encounters with the natural world motivate commitment to environmental ethical action, and environmental ethical actions yield spiritual experiences of transformation and renewal.

One dimension of a spirituality of environmental action is its service orientation. Restoration practice in particular is a type of service to others, to land and its (human and nonhuman) beings. As we have seen, some restorationists view nature as holding intrinsic, sacred value given its nonhumanly created status and self-renewing capacity. When nature is understood in this way, restoration becomes a form of service to the wild others in our midst, or a type of obeisance to or active recognition of creation's gracious abundance. Additionally, when ecosystems and human inhabitants together have been unjustly treated and dually impoverished, restoration activities constitute a type of joint service to land and people, giving back to both that which has been wrongly taken away.

Second, activities can be undertaken as ritual restorative action, in which certain practices, such as weeding and planting, become meditative, even prayerful habituated acts. For example, through the actual transformative practices of, say, ripping out buckthorn or putting in water bars or gathering seeds or working alongside fellow restorationists, a sense of clarity about oneself, others, and the surrounding world may arise; a deep conviction regarding the ultimate wrongness of particular injustices to land and its people may be formed, and profound insight into solidarity with other persons and creatures may be created. Experiences of transformation and renewal born of restoration practice are not only created through contemplative encounters with the natural world, but they also are fostered through the actual transformative ethical act of reversing "a tide of degradation."

Closely related to this is a third dimension of a spirituality of environmental action: it is an activity that connects the agent with the divine aspect or sacredness of creation. Religious restorationists often speak about restoration activity as a type of devotional practice that enables them to experience the natural world as in some sense transcendent of the human self. Benedictine restorationists, for example, refer to the ways in which nature's self-sustaining processes and patterns point to God's abundant provision and care for the creation. Native American restorationists view the Creator as earth's original restorer. Deep restorationists find in their work a witness to the table of nature's gracious abundance. Each of these perspectives refers to a dimension of the natural world that is beyond, more encompassing, or deeper than the individual human self.

Fourth, a spirituality of environmental action involves a revisioning and renewing of the self in community with others. This takes place within direct experiences of the natural world, and in relation to others who are likewise engaged in environmental ethical practices. As already stated, many restorationists experience an uplifting high through working restoratively with others and with the land. The myriad emotions that restoration work activates, positive and negative, can provide a basis for a deep psychological–spiritual connection to nature. Further, renewal of the individual self is often enhanced by the collective effort to bring about positive social ecological change in communities and society. Restoration practice is integrally connected to particular groups of people, and the broader environmental restoration movement, working collectively for the health of land and people. This becomes an especially significant dimension of restoration spiritual experience in Western industrialized societies where nature-based spiritualities, along with religious belief and spirituality in general, run the risk of becoming private, individually oriented therapeutic endeavors.[39]

Restoration spiritual experience is formed not only through personal encounters with the natural world; it is also formed through the experience of working collectively with small groups of people, and a larger movement, to create healthier ecosystems and communities. This notion is resonant with the notion of a spirituality of resistance put forward by some justice-oriented environmental authors (e.g., Roger Gottlieb, Val Plumwood), yet it differs from this idea in terms of how resistance is viewed. A spirituality of resistance, writes Gottlieb, "is about the attempt to bring into being a personal identity that takes past and present forms of social evil seriously and knows that it is living on an earth scarred by unjustified and irrevocable loss. As other spiritual paths may center on love of God or systematic prayer, on meditation or revelation, this one is bound by responsibility to protect the earth. In this path, spirituality and resistance to the destruction of life, human and nonhuman alike, are inextricably connected."[40] Similarly, a spirituality of environmental action shapes and is shaped by environmental ethical action in direct encounters with the natural world. It is also characterized by engagement with natural lands (damaged and healing) and community with others working for transformative social ecological change. Yet the idea of resistance in restoration practice is defined not in terms of earth's protection from destruction but in terms of earth's regeneration from degradation. I say more about this difference in chapter 5. Now, however, it is important that we further examine the claim that restoration activities, when understood as a type of spirituality of environmental action, can yield direct experiences of transformation and renewal within participants, for herein lies the heart of restoration-based spiritual experience.

RESTORATION SPIRITUAL EXPERIENCE

The heart of restoration spiritual experience is evidenced most clearly in the actual experiences of transformation and renewal that environmental restoration actions can engender.[41] As with restoration of land, the healing of individuals is central to restoration-inspired spiritual and moral experience. The restoration of what might be called the personal heart in this context is both a healing of human consciousness and a healing of the relationship between the self and nature.

In part, the notion of healing becomes central to the spiritual and moral experience of restorationists precisely because the nature that is the source of spiritual insight and meaning-making is a despoiled, degenerated, polluted, and disturbed nature—a nature in need of healing. The nature of the restorationist is not the nature of Muir's Sierra Nevadas or Sigurd Olson's northern Minnesota. And while we may agree in a general sense that the era of nature completely untouched by human activities has ended, the natural ecosystems that restorationists encounter and rely upon as a source of spiritual–moral experience have *really* almost ended.[42]

Impoverished soil, diminished biological diversity, damaged ecological integrity, degraded natural processes, these are the natural features of restoration-inspired spiritual experience.

It is also the case, nevertheless, that the nature that restorationists are healed by is also a nature that is, through assistance, coming back and recovering, slowly but surely. This raises an important question for accounts of nature spirituality: How does working with a wounded, yet healing earth shape the particular types of spiritual–moral experiences that are engendered? If nature in general is an orienting concept for nature spirituality, how does a damaged and recovering nature in particular matter for restoration spirituality—and for any nature-based spirituality that takes the current state of ecosystems seriously? Four key elements of restoration-based spiritual and moral experience offer a response to these questions. The first three focus on the personal relevance of a restoration spirituality, while the fourth emphasizes its communal relevance.

In the first place, restoration activity can enable a de-centering of the self through the realization that humans are dependent on and interdependent with larger nature. An important part of this de-centering is the recognition that humans are limited and finite in relation to the larger community of life. Restorationists perform their work with the dual, at times tension-ridden, realization that while humans are part of and play an active role within the natural order of creation, they are not the center, or even always comfortable members, of ecological systems or processes. Once again, as House reflects: "King salmon and I are together in the water.... It is a large experience, and it has never failed to contain these elements, at once separate and combined, empty-minded awe; an uneasiness about my own active role both as a person and as a creature of my species; and a looming existential dread that sometimes attains the physicality of a lump in the throat.[43]

In House's case the recognition of human limitation and a consequent de-centering of the self was formed through the active attempt to restore Chinook salmon to the Mattole River. Feelings of uneasiness and dread in relation to his own sense of selfhood and the human species more broadly characterized the de-centering that occurred when House stepped into the salmon's world. He writes later about the "hard knot of relationship in the act of killing a creature of another species." De-centering takes place in relation to the female salmon whose eggs are to be taken for fertilization, incubation, and future reintroduction into the river. The act of killing other creatures or parts of nature, and the "hard knot of relationship" that this produces, is a necessary component of any restoration work, as we have already seen. Burning grasses, pulling, cutting back, and removing invasive species, and so on involve troublesome elements of the human relationship with nature—ones that modern industrial culture, as well as the contemporary environmental movement, have not adequately dealt with, as restorationists often point out. Yet they are also, as both House and Jordan suggest, fundamental to our relearning "how to celebrate the true nature of the relationship [between humans and the natural world]."[44]

Paradoxically, while restoration yields a de-centered sense of self, it can also form an expanded sense of self in relation to nature. This too is a key element in restoration spiritual experience here. It can be understood in terms of Arne Naess's notion of the ecological self whereby people come to view and experience themselves as continuous, in some sense, with nature. This sense of self is nuanced in the practice of restoration, for restoration is an attempt to make way for ecological selfhood. The creation of a restorative ecological selfhood in the present requires a healing from the "scythe that went too far" in the degradation and destruction of ecosystems in the first place. The healing action comes not only in giving back to nature what was (in many instances, wrongly) taken, or in a type of redemptive penance, as Jordan describes it, for the expiation of ecological sin; but also, in turn, a sense of being forgiven by nature and a gratitude that says "thank you audaciously for the future" by making ecosystems as beautiful, healthy, and whole as we possibly can.[45]

The restorative ecological self, then, represents for the restorationist the search for wholeness in the midst of brokenness, an attempt to find, as Janisse Ray puts it, "wholeness in a fragmented land."[46] "This is the most urgent business we face in the twenty-first century, this question," writes Ray: "How, in all this fragmentation, can we lead lives of wholeness?"[47] For many restorationists, learning to lead lives in a context of wholeness means relearning, spiritually and ecologically, how to return to wildness.

This indicates another element of restoration-oriented spiritual experience. The return to wildness involves the restoration of a sense of wonder and amazement about earth's vast, intricate, prolific capacities. "It is time," writes Beatrice Briggs of the restoration group, Wild Onion Alliance, "for those who know the scientific story of the universe to lead us back into wonder, reverence and respect for what has been created over the past 15 billion years."[48] Or in the words of Mary Evelyn Tucker, what we need is a "restoration of wonder," a "reawakening in the human of a sense of awe and wonder regarding the beauty, complexity and mystery of life itself."[49] A restoration ethic is an ethics of wonder, proposes Daniel Spencer. By "immersing ourselves in our ecological homes, in gaining the intimate knowledge and familiarity of these homes that ecological restoration requires, we are opened anew to the wonder of this planet we call home."[50] "It is a wonder, a miracle, that all these prairie plants can actually grow up where there was once such degraded land," the restorationist marvels. Sister Walgenbach says, "The simple breaking open of a seed" that the restorationist experiences, "is witness to the miracle of life," it is "a reflection of the presence of God within creation."[51]

Finally, a key element of restoration spiritual experience is its communal dimension. That is, feelings of renewal, satisfaction, fulfillment, and hope in restoration work come not only working directly with nature but through working with others to heal the land. Despite tensions and challenges that arise within restoration work where multiple stakeholders are involved, a spirit of cooperation and mutual accountability in caring for particular natural places often grows in the activities of the work itself. Connections develop between people and particular landed places, and among people who work together to heal such places.

Consider the Acid Mine Drainage and Art (AMD&ART) project in Vintondale, Pennsylvania. The community-based AMD&ART effort has worked since 1994 to restore the Blacklick Creek and surrounding land. Here was an area that had become extremely polluted by acid mine drainage (AMD), a metals-laden, often acidic, orange, syrupy water that seeps and surges from abandoned coal mines throughout southwestern Pennsylvania and the Appalachian region. In Vintondale's case a former mine portal poured AMD into Blacklick Creek at a rate of two hundred gallons per minute.[52] The restoration work in this case occurred basically in the town dump, immersing the people of Vintondale in nature's underside, dredging up feelings of loss and lament, even a sense of hopelessness with regard to the postmining town and its creek and watershed's future.

Yet in the end, as well as throughout its process, the project was extremely successful in many ways. A series of passive treatment ponds and wetlands were created to transform the orange, syrupy liquid into clear, clean water. The AMD is continually piped from the mine opening across an old railroad bridge, passing through six successive pools that filter the polluted water. Pond 1 holds the piped AMD, and then ponds 2 through 4 serve as wetland treatment cells. Filled with plants and compost to slow the water and promote biological activity, they work to make the water less acidic by allowing the metals to settle out. Ponds 5 and 6 further clarify the water by raising its pH balance and stripping it of metals. By the time the former AMD-polluted water reaches the final pond, it has become clean and "legal" for other uses.

The ecological benefits of the AMD&ART project are significant, though its artistic, historical, and communal aspects are perhaps more so. As the water moves through the treatment system, it undergoes a visual transition as well. Changing from the red-orange AMD to the blue-green color of clean water, the flowing water is accompanied along the banks of the ponds by a Litmus Garden planted with thirteen native tree species, whose leaves seasonally change colors along with the filtering water. In autumn, the Litmus Garden trees turn deep red around Pond 1 and grade through orange and yellow to blue-green at the end of the treatment system in pond 6, creating a visual reflection of enhanced water quality. The Litmus Garden also serves as a focal point for a community fall celebration.

The clean water from the final treatment pond empties into a restored wetlands area (called History Wetlands), which now occupies the site of the original Vinton Colliery.[53] The former industrial site had been abandoned by the early

1980s and had since been covered with raw boney, or waste coal, four to eight feet thick. Seventy thousand tons of waste coal was removed and replaced with recycled soil, which blended a number of topsoil materials from combinations of bony residual, dredged material, wastepaper, yard waste, sawdust, mushroom compost, fresh cow manure, aged cow manure, processed cow manure, and chicken manure. More than ten thousand native wetlands plants were introduced, which has since provided habitat for many insect and bird species, including wood ducks, geese, and killdeer. Beaver, fox, deer, and other animals have also returned to the area, and ten bat boxes were erected with the hope of attracting native bats.

The site's history is most tangible in the wetlands area, where the remnants of the old colliery buildings rise from the wetlands as ghostly reminders in the landscape. "Our hope was to bring Vintondale's history back to the surface," states AMD&ART, "to celebrate both its proud past and its future commitment to environmental improvement at the same time."[54] Three public art installations—the Mine Portal, the Great Map, and the Clean Slate—lie within the History Wetlands and speak to the area's mining history and future of restorative engagement with the land.[55] A multipurpose recreational area with playing fields and picnic areas was also created on the restored site. Since then, the park has become "the new social center of Vintondale, bringing new pride and new activity to the community."[56]

The AMD&ART project was about more than just treating water; it was also about treating the whole community.[57] Throughout the entire process, from original design development to the present day, the community was deeply involved. This level of citizen involvement helped to generate broad public support, as well as a communal feeling to the park. It also helped to foster transgenerational relationships and experiences, where "the older generation is inspired to tell stories about the mining town's history, and the young listen, learning from the past."[58] Through trees that were planted, wetlands regenerated, passive ponds created, art celebrating the memory of local miners installed, fields for playing sport and picnicking implemented, shared feelings of renewal and hope came as the creek and surrounding land, and the community itself, began to heal.

Even as restorationists labor in the midst of degradation and fragmentation, they also, as just noted in the AMD&ART project, often describe deep communal experiences. "A special communion forms when people literally dig into the earth to reverse a tide of degradation, atone for past actions, seek a new way of relating to things other than human, or enjoy the pleasure of good company and good work."[59] Further, where restoration is of a landed place that holds particular meaning for those attempting to restore it, as was the case in Vintondale, people gain a deepened sense of belonging to their life-place. They are enabled "to attach to the space, embrace the spirit, and find personal meaning within that reciprocal relationship."[60] In this way, Leopold's dictum—to live with an ecological education is to live alone in a world of wounds—might be translated into: to live with an ecological restorationist's education is to choose to live with a group of people in a world of wounds in an attempt to heal them within and without.

TOWARD A CHRISTIAN SPIRITUALITY OF ENVIRONMENTAL ACTION

The more general spirituality of environmental action such as the one just described helpfully illuminates several retrievable elements in Christian (and also other) traditions. For there are resources within Christian versions of nature-based spirituality, as well as in other religious and secular spiritualities, that can also address the ways in which encounters with the natural world and with others shape spiritual and moral experience. Restoration practice, in other words, can serve as a guide, helping to flag ideas within Christian thought that may, in turn, beneficially inform understandings of restoration practice and spirituality. In this section, then, I note several ways in which a spirituality of restoration may elucidate a Christian spirituality of environmental action in particular. Following this, in the final section, I explicate a key Christian motif that I view as especially relevant to the further development of a restoration spirituality and ethic.

The elements identified above in general restoration spiritual experience raise important questions for Christian nature-based spiritualities. Most notably, restoration experience raises the issue of how engagement with damaged and healing nature should be considered within Christian ecological spiritual narratives. Christian accounts of nature-oriented spiritual experience have been based on broad generalizations rather than on specific, concrete, and vernacular environmental experiences.[61] For example, there has been a lack of attention paid to the ways in which this or that ecosystem may shape this or that type of spirituality. One wonders, for instance, how a forested landscape may form a forest spirituality; a prairie landscape, a prairie spirituality; a lakeshore, a lakeshore spirituality. Moreover, little attention has been paid to how these spiritualities converge and diverge based on various experiences of, say, divine presence, grace, forgiveness, healing, and salvation. Is the human experience of God's grace narrated differently, for instance, if living on the seashore of Martha's Vineyard or a farm in Iowa or a river in Louisiana?

Additionally, as the particular practice of restoration suggests, Christian nature–spiritualities have largely neglected to consider the ways in which concrete encounters with ecosytems as damaged and healing may shape particular dimensions of spiritual–moral experience. Ecological theologian Mark Wallace argues that a Christian nature spirituality calls religionists to pay particular attention to the neediest places and people within nature.[62] Yet his account raises additional, unanswered questions. For example, what actually happens within the recesses of the human heart in the Christian spiritual narrative when people encounter nature's degradation and healing? Do they, for instance, become awakened and enlivened by the wounded, though living, breathing Spirit moving within nature? Do they feel absence, loss, or impoverishment of sacred presence where earth and its beings have been damaged by past human activities? How do religionists speak about salvation, wholeness here on earth, even where land has become a degraded, polluted wasteland?

Christian restorationists are not without responses to questions such as these. For example, recall the ways in which the Benedictine nuns refer to the divine within the miracle of unfolding in the created order, even where order and integrity have been damaged by ill-conceived human activities. They also describe the ways in which the wasteland of the human heart is reflected in the wasteland of damaged ecosystems and, conversely, the ways in which the soul may be awakened to the divine in the process of working to heal wounded land. Sister Walgenbach refers, for instance, to the ways in which the filtering capacity of the (restored) prairie plants, with their seven- to eight-inch roots, testifies to the ways in which God cares for us and for the ongoing well-being of the land community. "The land does not ultimately belong to us, but to all the other beings residing within it—and to the next generation of human and non-human beings that will come to reside on it."[63]

General restoration perspectives also raise questions about the communal nature of restoration, questions that are important to Christians in particular ways. Again, responses are many. Ecological theologians beginning with Thomas Berry have used the idea of reinhabitation to refer to the integration of the human and biotic communities within particular ecosystems and the larger community of life.[64] More recently, religion and ecology scholar Sarah McFarland Taylor uses the concept of reinhabitation to describe the ways in which the "green sisters" in her study interpret their Roman Catholic spiritual ecological activity.[65] Similarly, Benedictine restorationists understand their restoration work as stemming from a Benedictine interpretation of reinhabitation. Drawing on the spiritual idea of stability, they view restoration work as integral to the idea of home-making and staying put within the religious, social, and ecological community of the monastery.[66]

The idea that God's saving grace is known through the life of a particular moral community has deep roots within the Jewish and Christian traditions. For example, in the Hebrew Bible God is thought to speak to and through the Israelite community. Later, New Testament texts narrate the ways in which the spirit of Christ is known within. Moreover, since ancient times Jewish and Christian theologians have thought long and hard about the ways in which God's forgiveness and grace is experienced through the faithful practices of these religious communities.

The question raised by restoration experience in relation to these accounts is: Might certain environmental practices such as restoration be understood as a type of faithful practice as well? Further, are there particular ways in which divine presence and grace are experienced when religious communities dig in their heels and stay put, choosing to reinhabit and restore land even where it has been damaged and degraded? In some Christian spiritual ecological views faithful environmental practice is understood as a type of stewardship, caring for a land that does not belong to humans, but to God.[67] In other conceptualizations it is viewed as a form of partnership, where Christians work in tandem with other living processes and beings to bring about a sort of holy democracy on earth.[68] In yet other Christian environmental spiritualities, faithful practice is narrated in terms of daily activities that awaken communities of people to the divine presence within creation.[69] Faithful environmental practice is understood as a way of fostering the reintegration of the human community "into the goodness of God's entire creation, that which God has pronounced *good.*"[70]

The final issue raised when restoration experience is considered within accounts of Christian nature spirituality relates to the intrinsically ambivalent dimensions of the human relationship to nonhuman nature. With few exceptions, recent renderings of nature-based Christian spirituality have overemphasized the positive aspects of the human relationship to nature, creating overly romantic accounts of the potential for humans to live harmoniously within their ecologies. For example, while McFague emphasizes the need for Christians to center on "healing the wounds of nature and feeding its starving creatures, even as they also focus on healing and feeding needy human beings," she nonetheless overlooks the ways in which such concrete encounters with "the needy" actually form certain dimensions of Christian nature spirituality.[71] In the end, it is the positive experiences and understandings of the divine in relation to the natural world that form McFague's sacramental spirituality. Humans can and should learn to "love nature by relating to it in the same basic way we relate to other people: that is, with respect and care. We see these earth others as we see the human others—as made in the *imago dei*—and therefore as both subjects in themselves and as imitations of God."[72]

Notable exceptions to this romantic bias in ecological theology include the work of Mark Wallace, H. Paul Santmire, and Catherine Keller.[73] Santmire's and Keller's accounts, however, focus on the issues of Christian liturgy and the doctrine of God, respectively, rather than on Christian nature spirituality as I am discussing here. Wallace focuses on Christian spirituality, although his perspective, in attempting to prove Christianity's potential for more pagan, earth-based understandings of the divine spirit within creation, leans toward the romantic.

The romantic focus in Christian accounts of nature–spirituality may be a side effect of the neglect of evolutionary theory's difficult aspects within ecological theology, as I have already noted. It also could be the product of environmentalism's romantic tendency in general when it comes to accounts of nature–spirituality.[74] Further, overly positive descriptions of Christian nature spirituality could stem from the fact that ecotheologians have paid little attention to the ways in which religious engagements with actual environmental social practices have worked to reflexively shape moral and spiritual experience.

Ecological restoration practice works to correct this bias, however, in its emphasis on the notion that "true community" involves a recognition and awareness of the intrinsically difficult aspects of the relationship between human and nonhuman beings and systems. For example, as noted above, Freeman House emphasizes the existential uneasiness and emotional ambivalence that often arise in restoration work where the irresolvable tension between being simultaneously a moral and biological animal is recognized. The constellation of emotions that often accompany acts of restoration is complex, making it difficult to separate them into discrete categories.[75] Given the extent to which most of us today are removed from the daily realities of nature's workings, developing a practical, spiritual response to the inherent dialectic of evolutionary biological life, and human life within it, poses one of the great contemporary challenges of our time. William Jordan has proposed ritual performance as a necessary component of productively confronting the ambivalent dimensions of the consumptive aspects of human life raised in restoration work. It may be that certain ritual dimensions of the Christian life may be interpreted through the lens of the restorative act as well.[76]

In the context of Christian worship, the making of rituals whereby the relationality of human and nonhuman life is affirmed and enacted, including the ambivalent dimensions of this relationality, may help form deeper, more reflexive meaning for human ecological acts such as restoration. Additionally, Christian sacramental rituals such as the eucharist—the ritual act of making communion (Greek: *eucharista*) between human and divine—might be "brought outside." When I lived in central Illinois, for instance, a group of restorationists happened, one spring, to arrange a burn of the elementary school's five-acre prairie plot on the Christian holy day of Maundy Thursday (the Thursday before Easter Sunday). While the day was set according to the season—an early spring prairie fire is most effective at controlling weeds in that it kills newly germinated weed seedlings and knocks back invasive cool season perennial weeds such as quackgrass, bluegrass, bromegrass, clovers—and the whims of the weather—it happened to be the only relatively still day of the two-week period we were hoping to burn—and it was a secular gathering at the public school grounds, the firing event nonetheless occurred on one of the holiest of Christian days.

When the small group of eight of us arrived that morning, the first thing we did was introduce ourselves, given that many of us had not met. We then did a clothing check to make sure that we were wearing garments made of all-natural (e.g., cotton) rather than synthetic (e.g. polyester) materials. Prairie fires can leap as high as twenty-five feet into the air, move as fast as six hundred feet a minute, and burn as hot as seven hundred degrees, causing synthetic material to quickly catch fire or melt into the skin. In large-scale burns, restorationists often wear protective suits with helmets and masks similar to those worn by firefighters. This was a small-scale burn, however, and thus a basic clothing check and safety reminder was sufficient.

The conservation practitioner then proceeded to go over the firing process from beginning to end. She first explained that the firebreak, a strip of mowed grass approximately ten feet wide around the perimeter of the prairie, had been watered thoroughly that morning, to best contain the fire. We would burn upwind, south to north this particular day, in order to avoid an uncontrollable burn. She would serve as the "fire boss," controlling the line of fine with a driptorch, a hand-held metal canister, often used in firing prairies, that is filled with a mixture of diesel fuel and gasoline and pointed downward to deliver the fuel (through a metal tube with a fire pad on the end) to the ground to start the fire. Two people would wear backpack-type water sprayers, filled with four gallons of water and weighing approximately thirty-five pounds, in order to control the fireline and extinguish any spot fires that may jump the line. One person would stand ready with the hose sprayer that ran off a water tank in the bed of the practitioner's pickup truck. Finally, we would need two "fire flappers" who would wield a long-handled tool with a large, black, duck-foot-shaped rubber pad at the end to put out small spot fires after the main line of fire passed.

Once we were clear about our jobs and tasks, we gathered our tools, made sure they were working properly, and got into position. The fire boss yelled "ready to fire" and leaned down to ignite the line of fire. Slowly working her way along the ground she continued to "burn in" the entire back firebreak until it had consumed the fuel within the first twenty to fifty feet and created a "blackline." She continued to lay down a fire line along both of the two edges of the prairie that were perpendicular to the original fire line. At this point, the fire began to gain momentum, bright orange flames leapt higher, and billows of black smoke thickened. The noise of burning plant debris heightened slowly and steadily like an oncoming freight train. A pheasant, then another, flew skyward. I shouted to my three small children to walk back to the school building. It was becoming extremely, frighteningly hot.

After all three sides of the prairie had burned in, the fireboss lit the remaining portion of the area, the windward side, creating a grand finale, a raging headfire, flames shot ten feet skyward, smoke centrifugally became dense. The flames advanced rapidly until they met the previously burned blackline, at which point they began to extinguish themselves for lack of fuel. The water sprayers and fire flappers walked slowly around the field, wetting edges and snuffing patches that were yet smoldering. The ground was smoking, black as night.

We chatted quietly while we put equipment away, congratulating each other on a successful burn. Upon leaving, the state conservation worker at the site shouted "Happy Easter" to all of us. And while this salutation was perhaps inappropriate at such a public event, it nonetheless made me see the restoration act with new eyes: that is, as a symbolic ecological revision and enactment of the eucharist, a public liturgy testifying to the sacramental, life-death-life, flourishing-suffering-flourishing process inherent to all of biological nature.

Even as restoration-based spiritual experience spotlights overlooked elements within Christian perspectives on spirituality, theological resources deep within the Christian tradition may help guide the ongoing development of a restoration spirituality and ethic. Here I propose one such resource: the concept of partnership. Perhaps more than any other theological motif, the idea of partnership relates most directly to the spirituality of restoration. I explicate it further here with the above restoration issues in mind.

A CHRISTIAN SPIRITUALITY OF LAND PARTNERSHIP

The idea that humans are and should act as partners with other creatures and land as a whole has a rich history in both conservation and Christian thought. Within American conservation thought, understandings of partnership can be traced to the mid-nineteenth century and the work of a small group of conservation thinkers and land managers. Most notably among them were Vermont statesman, and later US ambassador to Italy, George Perkins Marsh, who argued that humans must be coworkers with nature, aiding in "the reconstruction of the damaged fabric which the negligence or the wantonness of former lodgers has rendered untenantable."[77] Acting as restorative partners with damaged natural lands, according to Marsh, provided the corrective measure for creating healthier ecosystems and human–nature relations.

In the mid-twentieth century, conservationist Aldo Leopold further developed the idea of partnership between people and land, specifically in relation to the private landowner. People and land could partner together in a mutual, reciprocal, relationship, one that benefited them both, according to Leopold. Leopold wrote: "When land does well for its owner, and the owner does well by his land; when both end up better by reason of their partnership, we have conservation."[78]

In Christian thought, the notion of partnership has roots in the Hebrew Bible.[79] Ecological theologian H. Paul Santmire writes that a biblical view of partnership is characterized by three basic expressions: creative intervention in nature, sensitive care for nature, and awestruck contemplation of nature.[80] Interestingly, the practice of ecological restoration can be interpreted as representing each of these expressions: in restoration activities, humans may engage, for instance, in wild design (creative intervention), participate in hands-on regenerative tasks (sensitive care), and be caught up in the beauty of natural processes and the healing ways of land (awestruck contemplation).

As instructive as Santmire's interpretation is, it nevertheless fails to fully consider a view of partnership in relation to concrete environmental activities, particularly as a notion of partnership may provide a symbolic *and* practical alternative to stewardship models. Traditional theological understandings of land stewardship envision humans as trustees of a planet that belongs not to them, but to God. Views of partnership, on the other hand, de-center the place and role of humans in relation to earth and its nonhuman others, making them "plain member and citizen," to use Leopold's language, within the land community. Humans remain active, creative, and moral agents within theological understandings of partnership, as they are within stewardship views. Yet now humans are re-centered as partners with and participants among, rather than stewards over, the rest of creation, striving to listen to and live mutually with the nonhuman others in their midst.

Prominent among Christian authors representing a partnership view of humans and nature is twelfth-century monk and mystic Francis of Assisi.[81] As is by now well documented, historian Lynn White Jr. criticized the medieval Christian worldview as the "most anthropocentric the world has ever seen," proposing instead St. Francis's nonanthropocentric, earthy theological worldview. Yet, White (as with Santmire) neglected to consider how particular environmental practices might further illuminate and make concrete religious worldviews such as Francis's, limiting the work such narratives have come to perform within Christian environmental thought.

Francis envisioned a kind of holy friendship between people and nonhuman others, based on God's creation and call to human and nonhuman creatures to live together in ways that were mutually and reciprocally beneficial and fulfilling. In accord with this, Francis conversed and made pledges with various animals, including most notably a wolf that was terrorizing the village of Gubbio, and the birds that he preached to (and who preached to him) and called brother and sister.[82] St. Francis's spirituality and ethic of partnership between humans and life's other creatures also involved an earthy asceticism, the necessary underside of the Franciscan vision of joy and love. According to Francis, humans were enabled to become true partners with the suffering poor, human and nonhuman, by releasing attachment to worldly possessions and thus being freed to better hear the small, quiet voices within the world around. The human capacity to hear nature's voice, and to act in ways that fostered, rather than oppressed, other creatures' intrinsic potentiality went hand in hand with Francis's simple life of voluntary poverty. St. Francis's writings on poverty, suffering, and death provide a counterbalance to many otherwise overly romantic views of partnership between humans and wild others.[83]

Despite rich roots in historical traditions of spirituality and ethics, the notion of partnership between human and nonhuman creatures was for many centuries largely lost within the Christian tradition. This may have been an inadvertent oversight, perhaps resulting from the fact that Francis was a marginalized figure in Christianity, both in his day and historically, despite his patron saint status among contemporary environmentalists. But there may also be deeper theological reasons at play with partnership's neglect within Western Christianity and culture.

Traditional Christianity has long been suspicious, for example, of theological viewpoints that blur or dissolve distinctions between Creator and creation, heavenly and earthly, grace and nature, and human and nonhuman. As ecotheologians have pointed out, the idea that human beings are created uniquely in God's image to act as divine-like stewards or caretakers—and, on some interpretations (Gen. 1:26), to have dominion—of earth has served as the predominant environmental metaphor in Western Christianity.[84] The notion that human and nonhuman beings are created equally by God to live together in mutual and reciprocal partnership, on the other hand, calls into question traditionally held beliefs about the fundamental nature of individuals, other kinds, and the divine.

More recently, however, and with the rise of environmentalism, feminist philosophers and theologians have retrieved and recontoured theological metaphors of partnership. These thinkers, for example, have expanded upon notions of partnership to refer to the types of relationships between women and men and the earth that are created when people join with, rather than dominate, others in a shared struggle for justice and equality.[85] Feminist environmental historian Carolyn Merchant's notion of partnership, for instance, includes the dimensions of equity between men, women, and other creatures that feminist ecological theologians often stress as well. Partnership between human and biotic communities "entails a viable relationship between a human community and a nonhuman community in a particular place, a place in which connections to the larger world are recognized through economic and ecological exchanges."[86]

In addition to the social and ecological dimensions of partnership proposed by authors such as Merchant, a religious perspective on land partnership includes a consideration of the ways in which the sacred relates to the human–human relationship. A Christian spirituality and ethic of partnership in particular also involve the idea that religious communities have a particular role to play in the formation, and ongoing evaluation, of norms that constitute "good" forms of land partnership. Part of the purpose of this chapter, building on previous chapters, has been to provide the background for the initial development of such norms. For as I stated at the outset of this chapter, restoration practices require some kind of norms or criteria if they are to provide a basis for the generation of an environmental ethic and spirituality. With the background that this and other chapters have provided, I propose six precepts of a spirituality and an ethic of land partnership:

- Culture and nature are inherently, evolutionarily, and ecologically connected;
- Human and nonhuman beings are active subjects that rely on the health of the land community for their survival and flourishing;
- People and natural processes and systems can and should be interwoven in ways that promote both human and biotic health;
- Human–land partnerships are ongoing, dynamic, evolving encounters; over time, they are enacted variously;
- Human–land partnerships inevitably involve conflicts and ambiguities in need of addressing in order for the deepest type of relations to develop;
- Divine presence is experienced in various ways in and through partnering actively with natural systems and beings.

These precepts, as already noted, provide an initial framework that will require additional work, including input from religious communities engaged in partnering actions, in order for a full-fledged spirituality and ethic of partnership to be developed. Further, additional environmental practices and the spiritual–moral experiences they may yield will need to be evaluated. Accordingly, as such practices and experiences are explored, precepts of a land spirituality and ethic will need to be refined and reformulated. The spirituality and ethic of land partnership proposed here nevertheless provides an initial framework for understanding the types of relationships that may be required to generate experiences personal and communal transformation and renewal.

CONCLUSION

Restoration practice has a distinctive capacity for forming spiritual experience and meaning in relation to particular landed places. It can provide a context for the de-centering of the human self in relation to the biotic community of life. This sense of loss or minimization or relocating of the human self is often accompanied by feelings of uneasiness, limitation, and ambivalence related to being actively human within the larger evolutionary ecological world. In turn, humility in the face of a larger, gracious nature is often a mark of restorationists.

Even as the restorationist feels her inadequacy in relation to the magnitude of nature's healing power and evolutionary historical lineage, she is drawn into its inherently transformative, renewing ways. She recognizes the graciousness of nature's abundance. Divine care is witnessed in and through creation's underlying order and ways of providing for its beings. The restorationist experiences an enlargement of the self in relation to the landscape in which she works; alongside feelings of sorrow, lament, and even guilt and anger in relation to land's degradation, a sense of fulfillment, satisfaction, hope, amazement, and wonder at the healing capacities of land and the human spirit may arise. With others, the restorationist attempts to heal the wounds within and around. A sense of patience, cooperation, trust, and mutuality can be born among a group of people working to heal a damaged landed place.

Restoration-oriented spiritual experience helpfully highlights several elements in Christian accounts of nature-based spirituality. In particular, it suggests ways in which vernacular environmental practices may reflexively shape religious understandings of nature, community, self, and divine presence. Just as understandings of restoration practice are dynamic, reflecting the character of land itself, so too should be understandings of Christian nature—spirituality. There are, nonetheless, resources deep within the Christian tradition that may be drawn upon for addressing the ways in which direct encounters with the natural world shape spiritual experience. The theological theme of partnership provides one such example. As with perspectives on restoration, practice-oriented formed notions of Christian ecological spirituality should be clarified and refined, critiqued and reformulated, through processes of moral–communal discernment within and among Christian communities. Descriptions of nature-based spirituality and the ways in which the sacred is understood in relation to the unfolding order and dynamism of creation deserve similar attention and care.

A final remark is in order related to the ambivalence of spiritual and moral language within secular scientific restoration writing. As I have noted elsewhere, not all restorationists are as comfortable interpreting restoration in spiritual terms as the ones referenced here. Some restorationists, for example, state that they want their ecology straight up ("no nonscientific values with it, please") while others write that they see no need for the saving of their soul.[87] Both of these concerns, while legitimate, nevertheless overlook the potential of restoration to recover the study and application of ecology as a healing art, one that may help reconnect humans with ecosystems in ways that are more cooperative and durable. The issue of the connection among science, religion, spirituality, and ethics is one that pervades this project. More particularly, restoration practice raises the question and also sheds light on the ways in which spiritual values may and should be integrated with scientific and ecological ones. This issue is considered further in chapters 5 and 6. Prior to that, however, I turn to examine yet another dimension of restoration's potential as a healing art and science; that is, its potential for creating collective values in relation to particular landscapes and communities.

NOTES

1. For the full description of this project, see the Final Report of the "European Frogbit Spread Prevention Pilot Project—Year Three," November 11, 2011, Lewis Creek Association, http://lewiscreek.org (accessed April 14, 2012).
2. Lewis Creek Association, "About LCA," Lewis Creek Association, http://lewiscreek.org/about (accessed May 28, 2010).

3. Personal communication, May 7, 2008.
4. On Augustine's and Aquinas's understandings of nature, see Farley, "Religious Meanings," 109.
5. Calvin, *Institutes of the Christian Religion*. For an excellent treatment of Reformed nature–spirituality see Lane's *Ravished by Beauty*.
6. For more on the idea of sacramentalism, see the section on an "Intrinsic-sacred" dimension of a restoration meaning for nature in chapter 3.
7. See O'Brien, *Ethics of Biodiversity*, 59–60.
8. Ibid.
9. Ibid., 63.
10. Rasmussen, *Earth Community*, 239.
11. McFague, *Super Natural Christians*, 172.
12. Ibid., 174.
13. Rasmussen, *Earth Community*, 247, 239.
14. For this idea, I am indebted to Gottlieb's article, "Transcendence of Justice," 149–66.
15. Gottlieb, *A Greener Faith*, 157.
16. See, for example, Aristotle, *Nicomachean Ethics*.
17. See, for example, Augustine, *On Free Choice of the Will*.
18. See, for example, Day, *Loaves and Fishes*.
19. On this, see Plumwood, *Environmental Culture*, 218–21. Also see the various proposals related to building a new ecological consciousness in Kellert and Speth, *The Coming Transformation*.
20. See, for example, the collection of essays that speak variously to this research question in Kellert and Kahn's *Children and Nature*; Dunlap and Kellert, *Companions in Wonder*.
21. On the ways in which the ecological world shapes various stages of childhood development, see Heerwagen and Orians, "Ecological World of Children," in Kellert and Kahn's *Children and Nature*, 29–63. For a more narrative account of childhood experiences of nature, see Dunlap and Kellert, eds., *Companions in Wonder*.
22. Quoted in Norris, *Dakota*. This point relates to the current debate about how human activities and ecologies form the human mind. See, for example, Rolston, *Genes, Genesis, and God*.
23. Quoted in Van Wieren, "Children in the River," 287.
24. Emerson, "Nature," 382.
25. Ibid.
26. Ibid., 382–83.
27. Ibid., 397–400.
28. On this idea see Petersen, "Hunting for Spirituality," 195.
29. Ibid.
30. Quoted in Petersen. See Stange, *Woman the Hunter*.
31. Gottlieb, "The Transcendence of Justice," 157.
32. Ibid.
33. Ibid.

34. The exact quote is "Nature is loved by what is best in us." Emerson, "Nature," 387.

35. Ibid., 158.

36. For a related account of practice-oriented Christian spirituality, see Bass, *Practicing Our Faith*.

37. Wuthnow, *After Heaven*, 169.

38. Ibid., 198.

39. Of course, this is not inevitably the case when it comes to Western nature spirituality. For instance, deep ecological mysticism has historically been connected with the social movement of bioregionalism. Within contemporary religious environmentalism, examples of connecting eco-mystical experience and eco-political action may increasingly be cited. For examples of this, see Gottlieb, *A Greener Faith*. On the individual-therapeutic aspect of American religion, see Wuthnow, *After Heaven*, and Robert Bellah et al., *Habits of the Heart*.

40. See Gottlieb, *A Spirituality of Resistance*, 2.

41. For the discussion that follows, see my article in *Worldviews*, "Ecological Restoration as Public Spiritual Practice."

42. See McKibben, *The End of Nature*.

43. House, *Totem Salmon*, 13.

44. Ibid.

45. These two phrases, "the scythe that went too far," and "to say a thank you audaciously for the future," come from Cindy Goulder's poem "Volunteer Revegetation Saturday," cited in the preface.

46. Ray, *Pinhook*.

47. Ibid., 98.

48. Briggs, "Help Wanted," 24.

49. Tucker, *Worldly Wonder*, 52.

50. Spencer, "Restoring Earth, Restored to Earth," 431–32.

51. Personal communication, May 28, 2010.

52. Reece, "Reclaiming a Toxic Legacy," *Orion*, November–December 2007. The AMD treated in the AMD&ART project comes from Vintondale Mine No. 3, which was opened in 1899 by the Vinton Colliery Company (Vinton Colliery historically operated six mines in the Vintondale area, though now, all are abandoned, filled with water and the resulting AMD discharge). See the AMD&ART website, www.amdandart.org/projectindex.html for the description of the project cited here (accessed May 16, 2012).

53. The waste coal was removed through a GFCC (Government-Financed Construction Contract) Permit, issued by the Office of Surface Mining and the state mining office, that allowed a coal hauler to remove seventy thousand tons of material at no cost to AMD&ART. It then worked with Dr. Charles Lee (www.RSMT-LLC.info) and the US Army Corps of Engineers (Pittsburgh District and ERDC, Vicksburg, MS) to produce the recycled soil to cover the excavated site, again, at very small cost to AMD&ART. Other specifically blended soil was used to plant one thousand trees and shrubs that created the Litmus Garden, and for the recreational fields and other areas throughout the park. Ibid.

54. See the AMD&ART website, ibid.
55. The Great Map Project, for example, is a large mosaic inspired by the original 1923 Sanborn Insurance Map of the Vinton Colliery. It overlooks the wetlands, "linking the humanities and the arts in its visual and interpretive power. Standing upon the map, visitors can locate and compare the present landscape and buildings to those of the mining times. The eighteen-by-twenty-two-foot platform holds a nine-by-fifteen-foot mosaic and is located across from the former entrance to the Mine No. 6 Portal and immediately adjacent to the Ghost Town Rail Trail. Framing the map are 131 granite tiles, fifty-four of which are laser etched with community images, newspaper headlines, and text. The word 'hope' has been translated into twenty-six of the languages once spoken in Vintondale. Visitors and residents alike now have the opportunity to recognize, understand, and reflect upon Vintondale's rich community history, an area often overlooked in mining towns." Ibid.
56. Ibid.
57. Ibid.
58. Ibid.
59. Higgs, "What Is Good Ecological Restoration?," 342.
60. Thayer, *LifePlace*, 72.
61. See, for example, Sallie McFague's "horizontal sacramental spirituality in *Super Natural Christians*, Rosemary Radford Ruether's "ecological spirituality" in *Gaia and God*, and Rasmussen's "evolutionary sacramentalist cosmology" in *Earth Community, Earth Ethics*, each of which refers to a nature or earth writ large in conceptualizing a Christian nature-based spiritual account.
62. See Wallace, *Finding God in the Singing River*.
63. Sister Mary David Walgenbach, personal communication, May 28, 2010.
64. See Thomas Berry, *Dream of the Earth*, 163–179.
65. See Sarah McFarland Taylor, *Green Sisters*.
66. Sister Mary David Walgenbach, personal communication, May 28, 2010.
67. On this view, see, for example, the collection of essays on stewardship in Robert James Berry, *Environmental Stewardship*.
68. Rasmussen uses the term "holy democracy" to refer to St. Francis of Assissi's theology and spirituality of partnership with creation and its beings. See his *Earth Community*, 236.
69. See Sarah McFarland Taylor, *Green Sisters*.
70. Spencer, "Restoring Earth, Restored to Earth," 430.
71. McFague, *Super Natural Christians*, 4.
72. Ibid.
73. See Wallace's *Finding God in the Singing River*, Santmire's *Ritualizing Nature*, and Keller's *Face of the Deep*.
74. On this idea, see Jordan et al., "Foundations of Conduct."
75. Stephen R. Kellert has proposed a typology of nine dimensions along which humans may inherently connect with nonhuman nature (e.g., aesthetically, materially, scientifically, spiritually). See Kellert, *Kinship to Mastery*. Such fundamental emotional responses to the natural world, although intrinsic human capacities,

require additional learning opportunities in order to become functional. A critical element in this learning process must, in my view, involve opportunities for people to deal productively with the plural and at times conflicting feelings that often arise in relation to nature. I discuss this element in various places in this book.

76. Michael Northcott proposes that the activities of the parish church today can still provide a context for religionists to materially connect to local natural lands. Northcott's proposal echoes sixteenth-century Anglican theologian Richard Hooker's vision of the ritual, familial, moral, and productive role of the parish church in English society. See Northcott, "From Environmental U-topianism."

77. Marsh, *Man and Nature*, 35.

78. Leopold, "The Farmer as a Conservationist," 294–99, cited in Fader and Callicott, *The River of the Mother of God*, 235. Leopold wrote about a variety of activities that served to partner humans and land, including farming, hunting, and restoration practices.

79. See Santmire, "Partnership with Nature," 381–412.

80. Ibid.

81. White, "The Historical Roots of Our Ecological Crisis."

82. Brown, *The Little Flowers of Saint Francis*.

83. A good example of this is his "The Canticle of Brother Sun." The canticle famously begins by giving detailed praise to various creatures and aspects within the natural world. Yet it is largely overlooked that the canticle ends with praise "for our Sister Bodily Death / From whom no living man can escape!" Francis goes on to cite the Christian solution to bodily death in the eternal blessing of God in Jesus Christ, though the fact that he praises bodily death as an inherent given of created life is notable. Ibid.

84. On this, see Rasmussen, *Earth Community*, 228–36.

85. See, for example, Ruether's edited volume, *Women Healing Earth*.

86. Merchant, *Reinventing Eden*, 224.

87. On this point, see Meekison and Higgs, "The Rites of Spring," 73–81.

◦～◦

REGENERATING COMMUNITIES OF PLACE

Public Restoration Values

Those who labour in the earth are the chosen people of God,
if ever he had a chosen people, whose breasts he has made his
peculiar deposit for substantial and genuine virtue. It is the
focus in which he keeps alive that sacred fire, which otherwise
might escape from the earth.

—Thomas Jefferson

K ey ecological, spiritual, and moral dimensions and implications of restoration practice and experience have begun to be sketched in previous chapters. Beyond these, however, there are communal and social aspects in restoration practice that need to be explored. Restoration, as we have seen, is a group activity, often involving a cadre of practitioners—restoration ecologists, landscape architects, volunteers, land managers, farmers—working collectively to implement particular goals and objectives. And although restoration efforts today run the risk of technological drift—that is, of becoming a practice dominated by professional restoration firms and scientific experts—they nonetheless continue to involve considerable numbers of volunteers. As I stated in the introduction, more than any other environmental activity in the United States, restoration projects rank highest in number of volunteer hours logged.

In this chapter we continue to expand our understanding of the types of values that ecological restoration can generate within participants and society. In particular I examine some of the communal values restoration activities can generate in relation to particular communities of people. I propose that insofar as ecological restoration efforts are characterized by the ethical principles of recognition, participation, and empowerment, they are public participatory activities for the good of the community; they can, therefore, positively contribute to the formation of good ecological community. Further, I argue that participatory forms of ecological restoration can promote additional social ecological values that may create a deepened sense of place in relation to particular natural and cultural landscapes. In closing I provide two examples of ideal restorative communities of place based on the ethical norms proposed. I begin by critically examining some of the main ways in which restoration ethicists have understood the notion of good ecological community in relation to the practice of restoration.

PERSPECTIVES ON GOOD ECOLOGICAL COMMUNITY

Moral philosophers and theologians have thought long and hard about what constitutes good forms of community. They have wondered, for example, whether certain ethical norms, such as justice, should define democratic societies and, if so, what kind of justice and for whom. More specifically, foundational questions have been raised in regard to political liberalism and whether it overemphasizes the principles of individual autonomy and freedom to the neglect of collective empowerment and action. Further, ethicists have debated the positives and negatives of economic globalization and the ways in which it affects the formation of (or misformation) of local community.

Environmental ethicists too have examined issues related to the norms of good community. Here questions have centered on whether certain models of community might shape human patterns of action that are better or worse for promoting healthy natural environments. Some environmental ethicists have argued for a devolution, for example, of local forms of government, in the ways in which political and economic communities are organized, proposing instead forms of government and production that are "closer to the ground" and closer to home. Still others have argued for improved forms of civic environmental discourse and collective decision-making processes. And some have questioned the model of community altogether, arguing that principles of unity and harmony intrinsically minimize and homogenize difference—within human and biotic communities alike.

Restoration philosophers in particular have emphasized the ways in which good restoration inherently forms social as well as ecological values in relation to particular natural landscapes. Most notably, Andrew Light and Eric Higgs are once again in the forefront of thinking regarding restoration ecology and ethics. Light and Higgs have proposed that participatory forms of restoration can foster a democratic culture of nature, strengthening civic engagement in the environmental stewardship of local natural lands. Here I critically analyze and extend their theories of good restoration and community, based on the consideration of several justice-oriented concerns.

At the outset it may be helpful to recall the history behind Higgs's and Light's proposals regarding restoration's inherent democratic potential. Shortly after philosophers Robert Elliot and Eric Katz published their anti-restoration arguments in the 1980s and 1990s, respectively, Higgs and Light began developing counterproposals that focused on restoration's potential for promoting public, hands-on engagement with the care of local natural landscapes. Central to their arguments was the notion that good or beneficent restoration efforts would inherently involve social dimensions as well as ecological ones; that is, good restoration would foster not only healthier, more intact ecosystems (a point Elliot and Katz fundamentally questioned in relation to restoration) but healthier, more intact human relationships with those ecosystems as well. Fostering healthier human–nature relations, according to Higgs and Light, required that people were experientially engaged with their natural ecologies, something restoration activities were able to provide when public citizens were involved.

Since these early arguments Higgs and Light have developed additional proposals related to restoration's capacity for creating collective values in relation to land. Higgs, for example, uses the notion of focal practice in interpreting the ways in which restoration might generate cultural values in relation to particular ecosystems.[1] Following philosopher Albert Borgmann, Higgs views technology as giving shape to an overall pattern of life that alienates individuals from basic, daily nature-related practices.[2] Such practices, according to Higgs, are critically related to the formation of meaning in relation to individual and collective life. Focal restoration for Higgs serves as an antidote to technological restoration and the current technological drift that he sees within the ecological restoration movement. As an alternative, focal restoration promotes bodily and social engagement with particular natural places and, in turn, may become a form of meaning-making that is grounded in specific landscapes.

Focal restoration has two key aspects, according to Higgs: (1) participation—restoration as holding "inherent democratic capacity" in terms of its potential to promote "community engagement, experimentation, local autonomy, regional variation, and a level of creativity in working along with natural patterns and processes," and (2) celebration—restoration as promoting events that renew "the spirit of a community sharing in the regeneration of a native ecosystem" (e.g., celebratory burnings for tall grass prairie).[3] Further, Higgs proposes landscape coevolution—the idea that both ecological and cultural processes affect the evolution of natural landscapes—as the underlying model of focal restoration. Where human practices do not overwhelm the evolution of ecological processes, coevolution can be said to occur, proposes Higgs. This suggests that reflective learning takes place over time in relation to shifting cultural values and ecological circumstances.

Higgs graphically depicts this process of landscape coevolution as DNA-like strands of culture and ecology woven together over time: ecological history intersects with cultural memory, cultural reflection with historical reference conditions, and the ecological future with cultural imagination.[4] Focal restoration, argues Higgs, helps ensure that cultures and human activities do in fact remain or become tied to local evolutionary ecosystemic processes that encourage their ongoing self-renewing capacity and health.

More so than Higgs, Light's proposal related to restoration's potential to generate cultural values in relation to land emphasizes the civic political dimensions of restoration work. Recall that for Light, restoration activities can create moral-social value by promoting citizen participation in public environmental practices. Participation in restoration efforts, along with contributing to the good of the community, can lead for Light, as already noted, to the formation of more general moral environmental norms related to the stewardship of local natural lands. The creation of ecologically oriented, community-based moral norms has the potential to contribute positively to the development of "a new and more expansive 'culture of nature.'"[5] A culture of nature, proposes Light, is closely linked to a democratic model of participation, or ecological citizenship.[6] Ecological citizenship offers "the best model for achieving the full potential of restoration in moral and political terms." Light continues: "Our choice of how we shape practices and policies involving restoration is a critical test for how deep a commitment to encouraging democratic values we have in publicly accessible environmental practices."[7]

As we have already seen, the type of democratic ecological citizenship Light has in mind roughly follows "'classical republican' lines . . . which identify a range of obligations that people have to each other for the sake of the larger community in which they live."[8] Such might be called a form of ethical citizenship or citizenship as vocation, "where being a good citizen is conceived as a virtue met by active participation at some level of public affairs."[9] Light expands this notion of citizenship to include an environmental dimension. In sum: "if the point of ethical citizenship is to encourage people to take on responsibilities for each other in communities, then these responsibilities can be expanded to include environmental dimensions as well."[10]

Volunteer restorationists act as good ecological citizens, in other words, when they participate in the positive public-communal activities of restoration projects. Where restoration projects are only conducted by professional design and engineering firms without utilizing citizen volunteers, "an opportunity to foster such ecological citizenship would have been lost."[11] "When people participate in a volunteer restoration, they are doing something good for their community both by helping to deliver an ecosystem service and also by helping to pull together the civic fabric of their home," Light argues.[12] Similar to laws that exist to encourage voting, Light proposes that laws should be made that support volunteer participation in publicly funded restoration projects. Although the notions of focal restoration and ecological citizenship hold promising potential for an understanding of ecological community, there are nonetheless at least two inadequacies in these accounts. Both relate to concerns that are raised when the principle of justice is considered in relation to understandings of community and social relations. Since I want ultimately to emphasize justice concerns, it is useful to recap the meaning for justice I articulated in the introduction. This summary will be much too brief; nonetheless, the intent is to set the stage for my more-constructive proposals regarding the criticality of including justice-oriented concerns in a core understanding of good ecological community.

CONSIDERING JUSTICE

The notion of justice is generally understood in at least two ways in the environmental ethics literature. In the first meaning "justice" is applied in a particular way to individuals who have been marginalized as groups or in oppressed communities that experience, most acutely, a disproportionate level of environmental "bads" (e.g., pollution, contamination, desertification, deforestation). This is the predominant understanding of justice of the environmental justice movement.[13] According to this view the right to a healthy environment is a human and civil rights issue; where environmental injustices exist, they mirror present socioeconomic, racial, ethnic, and gender inequities. Here environmental justice refers to the demand for equity in the distribution of environmental burdens and benefits, as well as to participation, recognition, and empowerment, of the particular communities and individuals that are disproportionately negatively affected by environmental bads.

The second meaning of justice, recall, is used in relation to nature itself—the claims nature makes on humans. Nature has intrinsic value and dignity, and thus rights, that should be respected as such. Theorists vary in their specification of the parts of nature that have rights (and deserve justice): from all living beings, to sentient beings, to species, to ecosystems, to the whole earth as a living organic system.[14] Nevertheless, many agree that in general the requirements of justice, historically interpreted as giving to each person his or her own due, should be extended to nature. Although it is humans who are responsible to give justice, to human and nonhuman others, claims of justice, according to this perspective, are lodged both in human and nonhuman beings, and thus both deserve justice.

Despite the fact that the justice concerns are present to varying degrees in the ecological restoration movement, use of justice as a moral norm for evaluating the adequacy of good ecological restoration has received scant, if any, attention in the dominant ecological restoration literature. This brings us back to consider the justice-related oversights in Higgs's and Light's accounts of community-based restoration: first, both accounts use overly positive notions of community, over-looking issues of difference and pluralism; second, they also overlook issues of social justice in relation to participation in restoration efforts. I consider each of these concerns in the discussion that follows.

COMMUNITY, DIFFERENCE, AND PARTICIPATION

Ecological restoration thought, following a trajectory that runs through Aldo Leopold's land ethic and the modern-day conservation movement, emphasizes the model of community to represent ecological and human–land relations. The work of Higgs and Light, for example, well represents this view. Feminist philoso-phers and theologians, however, have critiqued recent notions of community (and associated "communitarian" theories) for the ways in which they homogenize dif-ference and otherness on the one hand and perpetuate oppressive social norms in the name of unity and similarity on the other hand. Further, these thinkers have raised questions regarding the tendency in dominant communitarian theories to place greater value on preserving the good and identity of the whole to the det-riment of individual subjects and groups. Moreover, appeals to community have been viewed as impractical on the one hand and utopian on the other.

Political philosopher Iris Marion Young, for example, critiques contempo-rary versions of communitarianism, or the view that the community is the ideal model for interpersonal and social relations (e.g., Alisdair MacIntyre and Michael Sandel), for their nostalgic tendencies.[15] Appeals to community as an antidote to liberal individualism (the notion that individual autonomy precedes sociality) according to Young, tend to be nostalgic and anti-urban. Echoing the Rousseauist romanticization of the ancient *polis* or Swiss *Burger* and its loathing of commerce, disorder, and the mass character of modern city life, communitarian proposals for social-political organization are wildly utopian, states Young.[16] Moreover, com-munitarian approaches, according to Young, tend to suppress difference among subjects and groups. "The impulse to community often coincides with a desire to preserve identity and in practice excludes others who threaten that sense of iden-tity."[17] This desire to maintain identity is expressed both in terms of the idea of a common consciousness or shared subjectivity, believes Young, as well as in terms of the concepts of mutuality and reciprocity. Both ways, "the ideal of community denies, devalues, or represses the ontological difference of subjects, and seeks to dissolve social inexhaustibility into the comfort of a self-enclosed whole."[18]

A Restored Notion of Ecological Community

Land-based communitarian proposals in particular are frequently accused of promoting the sort of wildly utopian, anti-urban bias that Young suggests. For instance, they are often viewed as overly romantic, representing a nostalgic and unattainable desire for a past way of life in which people lived more harmoniously, fruitfully, and happily with land. Further, as with political communitarian theories, ecological communitarian theories (e.g., J. Baird Callicott's "land ethic") have been critiqued for their "ecofascist" tendencies to privilege the good of the whole over that of the individual (human or nonhuman), and to overlook the question of whether the good of an individual ever competes with the good of the community.[19] In addition, ecological communitarian proposals have been accused of neglecting the issue of difference both between human and nonhuman and between diverse biological entities, as well as within the human community.[20]

One of the interesting aspects of the practice of ecological restoration, however, is that it is, at once, nostalgic and forward thinking while working squarely in the present. It is nostalgic for an ecological past where ecosystems were healthier and more vital, forward thinking about the ways in which land can and will repair itself if given the chance, and present-oriented in its doggedly hands-on commitment to work actively with the natural systems we have. Ecological restoration places a significant emphasis on history because it is oriented by the general idea that "we cannot know where we might and should 'go' ecologically speaking until we know where we came from." As environmental historian Marcus Hall writes, "Environmental restorationists see a better past alongside a worse present, but with a hopeful future. By acknowledging that time makes or breaks landscape, the restorationist uses history both to identify the need for restoration and to judge its success. Not only is history crucial to understanding restoration, history is crucial to restorationists."[21]

The restorationist's attention to history means that she values something that was once there, namely, the naturally evolving, intact ecosystem. In part, the commitment to assisting the recovery of the ecosystem's processes and functions and, hence the diversity of its organisms and species, represents a commitment to maintaining the land's health and dynamic stability. In part, however, it represents a commitment to continuing the narrative or story line of a particular place, including the specific attributes of its cultural and ecological character.

Narrative is based on experience, and experience is interpreted and communicated through the lens, language, and practices of particular cultures. The restorationists' interpretation of the past and desires for the future of natural landscapes, for example, are based on present perceptions and experience. This represents the idea that there is an inherent connection of past with future, "the idea that what *was* somehow flows into what *is* and then into what *will be*."[22] Nevertheless, this does not necessarily mean, as environmental philosopher John O'Neill describes, that past, present, and future are "linked by some single set of values which the present passes from past to future." Rather, values pass, according to O'Neill, "by argument both within generations and between them."[23] Moreover, cultural traditions and values can be internally and externally critiqued, and even transcended in some sense, though this too can serve the valuing of cultures.

The question we return to, then, is how, if at all, might and should understandings of nonhuman/nonhuman, human/nonhuman, and human-human difference be interpreted and narrated within accounts of "good" or "excellent" ecological community? Might such renderings reflect variety and differentiation as well as ecological wholeness and mutual exchange? Is it possible that restoration can promote both a greater shared sense of and connection to particular natural areas as well as social justice for all? How, if at all, might restoration address the question of whether the good of an individual (human or nonhuman) being ever competes with the good of the whole community?

Difference in Ecological Communities: Intra-Nonhuman Relations

Within restoration thought, considerations of difference and diversity are raised mostly within scientific ecological accounts of ecosystems. Restoration ecologists, while privileging notions of the whole—ecosystems, communities, populations, species—nonetheless recognize the ways in which biological diversity is integral and necessary to ecological integrity. Greater difficulty arises when ecologists attempt to discern just how much and what type of diversity is actually needed to maintain ecosystem stability and health.

This issue is evident in restoration ecological discussions about the value of non/indigenous and non/invasive species. Based on ecological restoration's emphasis on the regeneration of whole, historic ecosystems, species indigenous or native to the reference ecosystem are usually given preference over exotic or nonindigenous ones. According to the SER International Primer on Ecological Restoration, for example, restored ecosystems should "consist of indigenous species to the greatest practicable extent. In restored cultural ecosystems, allowances can be made for exotic domesticated species and noninvasive ruderal [plants that colonize disturbed sites] and segetal [plants that typically grow intermixed with crop species] species that presumably co-evolved with them."[24] So, in other words, based on this view, restored meadows in Europe might have some grazing livestock, and regenerated prairies might have some pheasant (naturalized in North America but indigenous to Asia)—but American bison should not be introduced into European meadows, and Scottish Highland mosses should not go into American savannas. This is because indigenous species, given their evolved capacity for adapting to the particular local conditions, are thought to have greater potential than exotic ones for promoting land stability and health.[25]

Some restorationists, nonetheless, believe that human beings should be less active in removing and promoting certain types of species, and instead allow noninvasive, nonindigenous species to remain in a restored ecosystem. Several groups of restorationists in Vermont, for example, told me that they were taking an explicitly "nonactive" approach to restoration, that is, they were focusing less on the active replanting and removing of certain species, and more on the reduction of negative human impacts to, say, natural buffer zones and soils. Vermont Family Forests (VFF), a sustainable forest restoration effort in central Vermont (see the end of this chapter for more on VFF), for example, focuses on the stabilization of water-bars in areas that have been intensively logged, so that soil can better hold to hillsides and existing indigenous saplings can better regenerate on their own.[26]

But perhaps the most heated restoration debates regarding the value of diverse biological entities center on conflicts that arise over sentient animals and indigenous plant species. In the famed Chicago Wilderness project, controversy erupted, for example, over whether the city's white-tail deer population should be controlled (killed, sterilized, or moved elsewhere) so that restored native trees, shrubs, and plants could survive (the deer had virtually destroyed the foliage in restored natural areas).[27] On the one hand, animal rights activists declared restorationists animal haters, as well as undercover agents attempting to manipulate natural areas without the public's consent. Restorationists, on the other hand, viewed the deer as overpopulated and domesticated pests, which were threatening the integrity of the community's natural areas. In the end, the deer population was controlled, through various methods, for the sake of the health of the native ecosystem as a whole.

Another controversial issue regarding indigenous species is the extirpation of certain indigenous species for the sake of regenerating or allowing the emergence of other indigenous varieties. In the article "Carving Up the Woods: Savanna Restoration in Northeastern Illinois," for instance, restoration ecologists critique the Cap Sauers Holding restoration project (Cook County, Illinois) of the Nature Conservancy's Palos/Sag Project for its heavily managed approach to the restoration of a savannah ecosystem.[28] These authors found that in some sites, white ash, basswood, black cherry, hawthorn, dogwood, and northern arrowwood, all native to northeastern Illinois woodlands, had been selectively cut and burned on "the assumption that these natives are 'weeds' in their own community and that their presence retards oak reproduction."[29] In this particular case, it was suggested that restorationists did not pay enough attention to the land's soil composition, Morley silt loam, which supports woodland in northeastern Illinois and southeastern Wisconsin, rather than savanna. With this ecological knowledge, the above authors argue, restorationists might not have made the mistake of killing so many indigenous tree species.

Both of the above issues raise important ethical questions regarding nonhuman difference for our discussion of "good" ecological community, especially whether there are circumstances when it may be morally justifiable for certain organisms or species, non/indigenous and/or non/sentient, to be eradicated for the sake of the whole ecosystem.

It may be that some general guidelines, such as the ones developed in the SER Primer, can be developed for evaluating the relationship between nonhuman beings and between individual organisms, species, and the restored ecosystem as a whole.[30] These may include, for example, the principles that (a) indigenous species of the historic ecosystem have precedence over nonindigenous species and individual organisms; (b) noninvasive (indigenous or nonindigenous) species have precedence over invasive ones, and; (c) sentient animals (domestic and wild) have precedence over nonsentient species or organisms, except where sentient animals are threatening destruction of an indigenous species in the historic ecosystem.

In most cases, however, restoration ecological and ethical issues such as these will need to be worked through on a case-by-case basis, with norms for "good" or "right" action determined collectively in accordance with a project's specific goals and objectives. I say more below about the importance of communal ethical decision making in fostering good ecological community. For now, we turn to the issue of difference between human beings in ecological communities.

Difference in Ecological Communities: Intrahuman Relations

Most notably, William Jordan is once again in the forefront of thinking regarding the topic of difference and community in ecological restoration thought. Jordan, as noted previously, critiques the environmental tradition beginning with Emerson and Thoreau and continuing through Leopold for its romantic view of community.[31] This critique is important and resonant with my argument. Nonetheless, Jordan's argument emphasizes the individual human emotional response to the encounter with an ontologically Other nature, overlooking important issues regarding difference within the human community, as well as regarding the relationship between the good of individuals and the good of the communal whole.

The oversight of difference with regard to social relations in dominant restoration thought is surprising, particularly when one considers the pluralistic character of the restoration movement itself. For instance, as restorationists are fond of pointing out, most public restoration projects involve scientists working alongside activists, land managers working alongside amateur gardeners, engineers working alongside artists, older adults working alongside younger ones. Restorationists also frequently emphasize the fact that restoration work crosses political, cultural, racial, and class lines, drawing people from various persuasions and groups to improve the health of local natural lands. From a multicultural and socioeconomic perspective, restoration projects can be found today in virtually every community globally, whether urban, suburban, or rural, black or white, rich or poor.

Further, decision making in relation to restoration projects is often characterized by a variety of differing perspectives on what land is for in the first place, what it should be restored to, and for what and whose purposes. Although restorationists tend to gain great satisfaction from the shared experience of digging into soil with the company of others, as Higgs emphasizes, the work itself, as well as the communal process of decision making that accompanies it, is often arduous. The restorationist's intimate, tactile participation with a particular ecosystem and group of people often makes him acutely aware of the multiple and at times conflicting dimensions of attempting, with others, to heal the wounds of land and spirit. Higgs's and Light's overemphasis on the positive dimensions of restoration collective experience not only misrepresents the variety of experiences inherent in the restoration movement, it obscures restoration's potential for creating deep forms of social relations, marked by variety and multiplicity, rather than only unity and commonality, in relation to particular people and natural landscapes.

Justice in Restoration Communal Decision Making

This leads to the second inadequacy in Higgs's and Light's proposals regarding restoration's community-forming potential. That is, not only do they offer overly positive notions of community, they neglect to consider processes of collective decision making that may in fact help safeguard pluralism and difference. For instance, although Higgs and Light both consider public participation as intrinsic to "good" restoration and "right" relationship between humans and natural systems, they nevertheless neglect to address what might make for good or right participation.

The issue of *who* participates and *how* decisions are made in relation to ecological restoration (and the sort of participatory democracy Higgs and Light support) raises several justice-related concerns. For example, both Higgs and Light leave out any discussion of the ways in which social injustices (e.g., the marginalization of historically oppressed groups in environmental decision making) may hinder the "inherent democratic capacity" of ecological restoration efforts and participatory democracy more generally. Higgs and Light both wrongly assume that focal practice or ecological citizenship, respectively, will by themselves promote the sort of participation that protects against social environmental injustice. Furthermore, they wrongly assume that the good of individuals will be fostered through the good of local political ecological community. Yet both of these assumptions overlook important issues of justice and difference that require attention in order for ecological restoration to positively contribute to the formation of good ecological community.

Christian environmental ethicist Michael Northcott, as noted at the outset of this book, issues a similar critique of dominant restoration philosophy in terms of its neglect of justice-related concerns. Echoing an early article (1978) by

Rosemary Radford Ruether, Northcott argues that the prophetic vision of restoration emphasizes the idea that social in/justice and ecological ill/health are integrally connected. The prophetic, ecojustice-oriented perspective of the Hebrew Bible (particularly the writings of the prophets, e.g., Isaiah), writes Ruether, provides a middle way between overly resource-based (stewardship) and romantic (nature mystic) approaches to the environmental predicament. According to this vision,

> nature no longer exists 'naturally,' for it has become part of the human social drama, interacting with humankind as a vehicle of historical judgment and a sign of historical hope. Humanity as a part of creation is not outside of nature but within it. But this is the case because nature itself is part of the covenant between God and creation. By this covenantal view, nature's responses to human use or abuse become an ethical sign. The erosion of the soil in areas that have been abused for their mineral wealth, the pollution of the air where poor people live, are not just facts of nature; what we have is an ethical judgment on the exploitation of natural resources by the rich at the expense of the poor. It is no accident that nature is most devastated where poor people live.[32]

Conversely, it is not only the degradation but the restoration of the oppressed poor and of the land that go hand in hand in the prophetic ecocultural vision, according to Ruether and Northcott.

Northcott focuses his argument for a prophetic ecocultural view of restoration on a restoration project in his home region near the Cairngorm plateau area of the Scottish Highlands.[33] This project is worth noting here, for although Northcott's justice-related view of restoration resonates with the one I propose, his perspective overlooks additional justice and restoration concerns. Northcott writes poetically at times about the beautiful, pristine natural landscape through which the famous Lairig Ghru wilderness walk (within the Highlands area) runs; he describes, for instance, the area's pure air, lichen-marked trees, dark mossy pine woods, broad and gracious valleys surrounded by striking rock formations and steep, grassy slopes, meandering river following through a glaciated heather-covered valley scattered with ancient Scots pine, and vast diversity of alpine plant and insect life.

Despite its wilderness-like natural landscape, however, Northcott points out that the Cairngorm plateau and the surrounding area have been shaped significantly by centuries of human inhabitation, including newer settlements by wealthy industrialists. Only recently, the National Trust for Scotland (similar to state land trust groups in the United States or the Nature Conservancy worldwide) purchased the Mar Lodge estate, which incorporates the Eastern side of the Cairngorm plateau and its approaches, with plans to "return this beautiful area to 'pristine' wilderness." "The human mark on the landscape is being systematically reduced, minimized, and the beauty of the wild 'restored,'" states Northcott.[34] Arguing that indigenous crofters inhabited this area of the Highlands until the mid-nineteenth century, Northcott observes that crofter farms and communities "were destroyed to make way for the vast houses and sporting estates of the would-be aristocratic lairds of modern Scotland after the Clearances."[35] The restoration project of the National Trust aims to "repristinate" the area as if humans, indigenous or modern, were never there.

Prophetic views of ecological restoration, according to Northcott, must be sharply distinguished from, as well as favored over, contemporary understandings of restoration. For the prophetic vision of restoration emphasizes God's redemptive work on behalf of the landless and oppressed poor, argues Northcott, rather than the return of wilderness to an original, pristine condition whereby "the right species once again can find their ecological niches."[36] Prophetic restoration is not concerned with a return to Eden, Northcott suggests. Rather, it is a witness to a "'new creation' characterized by justice and shalom."[37] Where "efforts to restore Highland land to the indigenous people go hand-in-hand with projects of ecological restoration," writes Northcott, we "may say that such projects do indeed come close to a biblical vision of ecological restoration."[38] Where they do not, "these projects fall far short of a biblical vision of ecological restoration."[39]

Northcott's critique, I realize, is based on the case of a particular restoration effort, and his concerns about it, and the lack of justice considerations within dominant contemporary restoration thought, are well founded. Nevertheless, Northcott's type of prophetic viewpoint leaves little room for additional conceptions of "good" ecological restoration. Furthermore, it fails to consider contemporary restoration work that addresses issues of human in/justice in relation to the degradation/regeneration of ecosystems. For example, it overlooks indigenous restoration efforts in which the regeneration of historic ecosystems, including wilderness-like natural areas, does in fact go hand in hand with the restoration of indigenous cultural practices and traditions.

Additionally, Northcott's conception of prophetic restoration overlooks important further dimensions related to the notion of justice. In emphasizing the importance of redistributive justice—giving the land back to the indigenous people to whom it rightfully belongs, in Northcott's view—Northcott neglects, for instance, to consider wider justice issues related to the concrete realities of environmental restoration decision making and dynamics. There are certainly reasons to focus on the restoration of land to indigenous peoples where it has been wrongly taken through forms of coercive and oppressive power arrangements. Yet it is also the case that "the scope of justice is wider than distributive issues," as Young and others (e.g., Michael Walzer and Charles Taylor) have persuasively argued in relation to modern political liberal thought.[40] In relation to restoration efforts, especially insofar as they are understood as experiments attempting to connect human and natural processes in ongoing, evolutionary, dynamic ways, matters of social justice become more significantly reflected in the actual dynamics and processes of environmental decision making. Further, emphasizing distributive forms of justice as Northcott does obscures the institutional forms of oppression of people and of land that Northcott argues caused the infraction against indigenous Highlanders in the first place.

The more relevant justice-related questions from an environmental restoration perspective relate to the three ethical principles noted in the beginning of this chapter: recognition, participation, and empowerment. In the context of the Cairngorm, these principles necessitate the following questions: Which human and nonhuman beings are acknowledged as the significant stakeholders in the project (recognition of difference)? Who made the decisions about the project in the first place (participation of stakeholders)? Why were not indigenous people at the table, and what is the best way forward—for the health of land and people now residing in the region (empowerment of marginalized groups)?

Recognition of difference, human/human and human/nonhuman, alone does not protect against the perpetuation of unjustified power imbalances and subsequent injustices in relation to the development, implementation, and oversight of environmental practices. Maximum public participation, and recognition of the "voice" of nonhuman nature, in the processes and procedures of environmental decision making, planning, and execution of projects is a requirement of justice itself. Justice understood dually in terms of recognition of difference and participation of affected stakeholders challenges dominant understandings of ecological restoration practice, asking: What human and nonhuman "voices" are not at the table and why? Are there ways in which decision-making processes and procedures are developed and enacted that marginalize certain individuals, groups, and biological entities and communities? Who decides in the first place who will participate, and according to what criteria? And how and why is this or that ecosystem or particular area of land chosen for restoration in the first place? To these ends, careful and sustained attention to the *process* of (and procedures related to) assisting the recovery of ecosystems is required when justice concerns are considered.

The above justice-related issues will need to be considered in determining what constitutes "good" restoration activity. Nevertheless, the ethical principle of human justice, whether distributive, procedural, participative, or compensatory, is not the only or final measure of good ecological community. There are additional social ecological values that characterize virtuous restorative communities and societies. I say more in the following chapter regarding justice concerns as they relate to broader cultural meanings for restoration. For now I further examine additional communal-social values that restoration can foster in relation to groups of people working collectively to heal damaged natural ecosystems.

VALUES OF RESTORATIVE COMMUNITIES OF PLACE

Contemporary political philosophers and theologians have tended to argue strongly in favor of either political liberalism and its ideal of individual autonomy on the one hand (e.g., John Rawls), or communitarianism and its ideal of social identity on the other (e.g., Alasdair MacIntyre, Stanley Hauerwas). Some theorists (e.g., Iris Marion Young, Michael Walzer, David Hollenbach), however, have attempted to hold onto, and merge, the best of both approaches, offering communitarian-leaning perspectives that still emphasize the importance of both individual and group identity and corresponding justice-related concerns. For example, Young proposes an ideal of social relations and politics, which begins from the "positive experience of city life."[41]

Environmental theorists too have offered proposals for linking notions of differentiation and community, and emphasizing heterogeneity over homogeneity. For example, as we have seen, bioregional writers have proposed the concept of reinhabitation as a model for human and ecological relations. Here the orienting concept of the bioregion is both cultural and ecological, rather than primarily social and political (as Young's "city life"). Further the emphasis on the principle of "self-realization" or "self-fulfillment" in bioregional thought takes account of both the well-being of individuals and the health of the communal whole. Ecological theologian Thomas Berry, for example, proposes six functions of a bioregion—self-propagation, self-nourishment, self-education, self-governance, self-healing, and self-fulfilling—each of which considers the fulfillment of both individual human and nonhuman members as well as the thriving of the whole community of life.[42]

Given its inclusion of ecological components, the bioregion, or "life-place," as landscape architect Robert Thayer defines it, includes yet extends the parameter and variety represented within city life. A life-place, according to Thayer, is "definable by natural (rather than political) boundaries with a geographic, climatic, hydrological, and ecological character capable of supporting human and non-human living communities. Bioregions can be variously defined by the geography of watersheds, similar plant and animal ecosystems, and related, identifiable landforms (e.g., particular mountain ranges, prairies, or coastal zones) and by unique human cultures that grow from natural limits and potentials of the region."[43]

Reinhabiting one's bioregion or life-place "means learning to live-in-place in an area that has been disrupted and injured through past exploitation. . . . It means understanding activities and evolving social behavior that will enrich the life of that place, restore its life-supporting systems, and establish an ecologically and socially sustainable pattern of existence within it."[44] It involves the active attempt to arrange human life in relation to ecological life in ways that are restorative,

127

fulfilling, equitable, and just, allowing every human and nonhuman creature to become "fully alive" within her life-place.

With this view of bioregional community in mind, as well as the above discussion on justice, I propose a normative ideal of restorative communities of place and the types of values that follow. Ecological restoration, as already indicated, is an activity that can foster deep connections and meaning between groups of people and particular natural landscapes. Here I propose six communal values that the practice of restoration ideally may generate.

Before doing this, however, I need to say more about what I mean by "restorative communities of place." The underlying assumption of the ideal of restorative communities of place is that strengthening ties to particular places is in fact a good and critical dimension of human life. For centuries, psychologists and theologians have written about the ways in which the need to develop a sense of belonging in relation to one's surrounding world is inherent to human development and fulfillment. For example, the mid-twentieth-century psychologist Erich Fromm famously posited "rootedness," or a sense of belonging, as an intrinsic human need.[45] More recently, scholars of community development and place have documented the ways in which ties to specific places and locales may contribute to an increased sense of human health, well-being, and fulfillment.[46] Arguing that we are social and ecological beings, with the need for intellectual and physical, spiritual and aesthetic stimulation and affiliations, these authors argue, for example, that human satisfaction, meaning, and fulfillment grow out of our connections with real places—cultural and natural.

I define restorative communities of place as a form of social and ecological relations, which reconnects groups of people and landscapes in particular locales. By particular locales I mean life-places or bioregions and the local communities and larger cities within bioregions. The scale of communities of place I have in mind is that level where it is possible to work concretely and vernacularly in ways that are regenerative for both people and natural landscapes. Regenerative activities can take on a variety of forms, including restorative design of the built environment, sustainable forestry and farming techniques, and ecological restoration practice. Restorative practices can occur in rural, urban, suburban, wild settings; communities of place can be largely natural environments or natural and built environments.

Communities of place imply the side-by-side living together of human and nonhuman creatures.[47] Particular places are significant and have specificity for the people and creatures living in them. Communities of place have objective and subjective qualities; they are defined by concrete built and natural surroundings, as well as personal and social meanings and memories that people associate with these surroundings. Environmental psychologists Setha Low and Irwin Altman write, for instance, that place "refers to space that has been given meaning through personal, group, or cultural processes."[48] Sometimes particular places take on sacred meaning; at other times, they simply elicit good or familiar feelings.

Communities of place are marked by specific sights, smells, sounds, and textures. The land may be flat and farmed, mountainous and forested, paved and built upon. One may smell cook fires, damp woods, saltwater, frozen air, or exhaust from cars and buses. Communities of place may be loud with traffic or farm machinery or the sound of birds or crashing waves. The air may feel gritty or pure, humid or dry; the soil may be sandy or clay-like; the trees and plants may be prickly or full of bright flowers.

Particular communities of place have variety, whether represented within the space of a local park, backyard, beach, stream, or vacant lot. Real places provide opportunities for real people, friends and strangers, to meet in ways that make us realize the astonishing plurality that exists even within our own species. Even with the homogenizing tendency of globalization, the diversity and changing face of a community's built environment—its schools, libraries, parks, restaurants, grocery stores, neighborhoods, roads, street signs, churches, farms, and city lots—provides a window into the diverse and changing ways people, over time, have chosen to arrange their lives.

Even as communities of place offer a context for exploring a richness of experience and beings, perspectives and social and ecological dynamics, they also provide a forum for fostering social relations and positive interaction. They are defined by commonality and similarity, and the coming together in shared spaces and public forums. But they are also places where we encounter strangers and acquaintances, coworkers and coreligionists, people from our own cultural heritage and those coming from different backgrounds, friends and perhaps family.

Communities of place are interconnected in physical, social, and symbolic ways even in today's global economy. These connections can strengthen communities of place in terms of creating positive exchange of resources and information. Physical, social, and symbolic connections of communities of places can also be weakened where patterns of production and consumption rely increasingly less on the particular skills and goods of the particular region. Additionally, many of us in today's mobile societies have become "place grazers," as landscape architect Timothy Beatley writes. "This modern phenomenon of multiple place experiences is both a vice and a virtue. On the negative side, our tendency to be 'place grazers' may serve to diffuse or dilute the commitments we feel to a specific place.... On the other hand, a variety of place commitments are helpful and even healthy, even to those places we visit infrequently."[49]

With this general understanding of communities of place in mind, I extrapolate from restoration experience six values required and fostered in restorative communities of place. The values are (1) cohabitation, (2) accommodating variety,

(3) promoting wildness, (4) sensuousness, (5) creating publicity, and (6) celebrating. I will address them singularly and together.

Cohabitation

The recognition of and awareness of difference between humans and nonhumans, individual and individual, and individual and groups of people is one that restorationists often recognize only implicitly. As already noted in previous chapters, restoration activities often accentuate the inherently ambivalent dimensions of being an active, participatory member of the human species within the naturally evolving biological world. We are in a sense "misfits" within ecosystems, as religious ethicist Margaret Farley writes, and restoration activities highlight this in ways that are at times acutely painful.[50] The removing, ripping out, burning, killing reminds the restorationist of this in the present moment. Further, the need for restoration in the first place is a reminder that humans of a generation or two before us also struggled with how to live as "misfits" within their ecologies. Even traditional societies, often touted for their capacity to live more harmoniously with land, often developed and enacted elaborate ritual practices to help the community productively deal with and regulate the relationship between human and nonhuman species. Rather than minimizing or overemphasizing the ontological difference that exists between ourselves and the wild beings in our midst, restorative communities of place express difference as cohabitation, a "side-by-side particularity," to use Young's term, "neither reducible to identity nor completely other."[51]

Additionally, restorative communities of place lead to differentiation of social groups and individuals. Restoration projects, for example, often draw in and affirm participation from a variety of individuals and social groups. Though community life as we now experience it often has many boundaries and forms of exclusion, restorative communities of place hint at what differentiation without exclusion may look like. Many community-based restoration projects, for instance, attempt to represent and involve the spectrum of residents living within the land area or watershed. And participants frequently cross distinct land areas and neighborhoods, farms or towns, in conducting the business and politics of restoration projects. This is not to say that restorative communities transcend the realities of struggle and difficulty involved with the attempt to cross and open social and ecological borders. Quite the opposite, the attempt to cooperatively cohabit particular landscapes with other people and creatures often accentuates differences in cultural attitudes regarding notions of ideal and damaged land. In the midst of such differences, restorative communities of place "engage the challenge of building a community of place that includes each creature and every human neighbor."[52] In the normative ideal of restorative community, borders are flexible and open, human and nonhuman beings are viewed and treated as side-by-side cohabitants.

Accommodating Variety

The interaction of groups of people and other species and ecosystems in restoration projects occurs partly because of the diversity of activities, interactions, and interests utilized in place-based restoration efforts. What makes restoration efforts inviting, draws various people, and gives them satisfaction is that they offer myriad types of work, and interactions with human and nonhuman beings, over the course of the restoration process. Regarding one category of work, William Jordan writes that restoration, besides providing a way of doing ecological research and being a kind of agriculture, involves the basic human activities of hunting and gathering.[53] Land needs to be cleared, seeds collected, handled, dried, sorted and stored, seedlings started and planted, others cut back or pulled, invasives removed, bundled, and burned, certain populations of species need to be controlled in the absence of their natural predators.

Additionally, restoration efforts require a variety of organizational skills for planning, implementing and leading. For example, phone calls often need to be made, emails written, agencies solicited, and other community organizations invited to participate. Both physical and social activities involved with restoration work are at times difficult and demanding as well as inspiring and rejuvenating. There is something for everyone in communities of place engaged in restoration work, and this variety of opportunity draws people out and strengthens their commitment to their natural areas. More fundamentally, restorative communities of place tap into and foster a variety of inherent human needs in relation to the natural world and its beings. Restoration work, for instance, promotes the human need to connect with nature aesthetically, spiritually, scientifically, and naturalistically.[54] Restorative communities of place engage us with the wild, nonhuman dimensions of our world by providing multiple, varied ways for us to connect in tangible ways with its beings, processes and ways.

Promoting Wildness

Most simply, wildness, as we have seen, is that quality of life (human and nonhuman) that is ultimately beyond human control. The term's etymology is from the Old English *wildeornes*, which in turn derives from *wildeor* meaning wild beast (wild + deor = beast, deer). Restorative communities of place, whether urban, rural, suburban, or wild, engage and foster the wild dimension of life, assisting the recovery of processes and functions that we ourselves did not create.

Historically within the West, the quality of wildness has elicited a sense of fearfulness, for it holds the possibility that one will lose control or be forced to lose control rationally and/or physically. Even as the dimension of wildness often represents fearfulness, people remain nevertheless intrigued by, drawn into, and take pleasure in the encounter with human and nonhuman life that are, in some sense, beyond human contrivance or mastery. For example, in encountering the intrinsic wildness of a simple seed, we may be struck by the fact that it, on its own, actually breaks open and grows. We encounter another species and may be astounded by the grace, ease, and stamina with which they move through the world. We may be surprised by the intricacy and beauty of the inside of a tiny wildflower. Restoration activities provide a context for people to learn from the other wild beings and processes that shape their own and other creatures' lives.

The dimension of wildness also may be said to exist within social relations that characterize restorative communities of place. There is at times a quality of relationality among people and between people and ecosystems that occurs that we ourselves do not control. In restorative communities, people come together to take action in ways that are at times wildly unpredictable and surprising, conflicting and cooperative. Restorative communities of place engage in processes of decision making in ways that are open to unpredictability, to learning from each and every person and creature, and to the encounter with wildness that may occur through those interactions. Restorative communities of place, as restoration projects, represent dynamic, unfolding communities modeled upon the organic wild character of evolutionary biological life and its processes.

Sensuousness

Restorative communities also foster difference, variety, and wildness in the sensuousness of interacting tactilely with particular places. As the literature on place suggests, positive connections to particular places, natural and built, involve a sensory component.[55] We become attached to, and in turn care for, places in and through the myriad sensory ways we are attracted to and connect with them. Particular places smell, taste, sound, feel a certain way. These sensory encounters form our experience and memory, sense of meaning and belonging in relation to a place. For example, we may recall fondly the smell of certain seasons, the taste of special foods, the feel of the ground, and the sounds of the city streets, forest, or fields.

Restoration work draws one out of her own self and forms place attachments precisely through the ways in which one is touched by the sights, smells, tastes and textures of the particular natural area being restored. The seeds may feel dry and smooth, the soil may smell damp and musty, the air may taste like fire, the birds may sound like phoe-be, phoe-be, phoe-be. Restorationists feel the tools in their hands, and the way they dig into and turn the earth or cut brush or spread seed. They taste coffee in the morning before a workday begins and sweat and dirt at the end of it. They hear the grunts and shouts and silence of their fellow workers. Ecological restoration work has an "eros of connection, of the physical," writes Mills.[56] Restorative communities of place embody tangible, vernacular encounters among people, creatures, waterways, valleys, creeks, beaches, mountains, air, sun, soils that form the bodily, physical makeup of a particular locale.

Creating Publicity

In political philosophy and theology, the public is often construed as a place of unity and common understanding, or at least a place where mutual, reciprocal dialogue or political adjudication is possible. Yet this is not always how we actually experience public spaces and dialogue. "Because by definition a public space is a place accessible to anyone, where anyone can participate and witness, in entering the public one always risks encounter with those who are different, those who identify with different groups and have different opinions or different forms of life."[57] This diversity is often apparent in restoration work, helping account for its energy and vitality. Restoration efforts provide important public venues, inside and outside, where people work together, interact, mingle, or merely witness one another, without necessarily becoming wholly unified in thought and action. Additionally, other species and organisms are included in the public natural spaces of restorative communities. Human and nonhuman beings, cultural and natural processes are interfused in ways that overlap and intertwine, though do not completely fuse in the public ecologies of restorative communities.

"Politics, the critical activity of raising issues and deciding how institutional and social relations should be organized, crucially depends on the existence of spaces and forums to which everyone has access."[58] In the public spaces where restoration projects are discussed and debated, people encounter other people with experiences, perspectives, goals, objectives, and desires that are different from their own. Where the land under consideration has multiple uses, as it most often does, the variety of viewpoints multiplies, at times causing deep divisions and "camps" where concrete decisions must be made. Restorative communities of place include public forums where the variety of issues, demands, and desires related to land use are expressed and heard.

Because restorative communities of place are a being-together of human and nonhuman beings, overlapping and differentiated residents, the requirements of social justice necessitate a politics of maximum recognition and participation. Restoration projects characterized by a politics of difference provide the ideological and organizational means for recognizing and affirming the variety of groups of people and species represented in the ecosystem. Diverse social groups are recognized through their political participation in decision making related to the restoration of the natural landscape, and by celebrating and affirming their cultural distinctiveness and contribution to the heritage of the region. Recognizing the "voices" of other humans and other species within the public ecological spaces of restorative communities of place means that consideration of nonhuman interests is weighed equitably with those of human interests.

In the normative ideal of restorative communities of place, community is not understood fundamentally as transcending or unifying differences, but as the place where variety is acknowledged and appreciated for its distinctiveness and richness. Echoing the character of ecosystems themselves, mutuality, reciprocity, and cooperation are affirmed alongside recognition and celebration of difference, variety, and particularity. Public partnership between human and nonhuman beings and among people is desired in restorative communities of place insofar as distinctiveness remains unassimilated and specificity is celebrated.

Celebration

Thomas Berry writes, "In conscious celebration of the numinous mystery of the universe expressed in the unique qualities of each regional community, the human fulfills its own special role."[59] Restorative communities of place involve ways for people to celebrate its unique cultural and ecological components, as well as the cycles of the place's seasons and processes of healing and renewal. Visual art, religious liturgies, song, dance, music, theater, communal festivals all are utilized in forming a place's cultural identity in relation to the restoration of the natural landscape.

Restoration efforts that help renew a spirit of place within the local community often involve artistic dimensions. A restoration project in Vintondale, Pennsylvania, a former mining town with a heavily polluted watershed, for instance, utilizes its restored wetlands and park area to host an annual community day and celebration. A Lutheran liberal arts college in Northfield, Minnesota, St. Olaf, brings prayer and liturgy out into fields of 150 acres of prairie that has been restored to the campus' formerly farmed surrounding natural lands. Benedictine restorationists in Madison, Wisconsin, include prayer, poetry reading, and a meal in their seasonal workdays, adding a celebratory, festive dimension to their ecological work. Celebration within restorative communities of place involves both a purifying and healing aspect as well as a renewing and fulfilling one. It acknowledges the intrinsic ambivalence and challenges involved in attempting to live in community with other human and nonhuman beings, as well as the deep values of communion, beauty, and even love that may be born where people become fully engaged with and enlivened by the larger community of life.

REALIZING RESTORATIVE COMMUNITIES OF PLACE

The ideal of restorative community is already being realized in varying degrees and in multiple ways in hundreds, even thousands of communities throughout North America and worldwide. The six values of restorative communities proposed above—cohabitation, variety, wildness, sensuousness, publicity, and celebration—are grounded in and shaped by these positive experiences of communities of place that already exist in some form. In addition, positive experiences of restorative community are yielded in and through collective decision-making processes that uphold the principles of recognition, participation, and empowerment of marginalized groups. In closing, I give two paradigmatic examples, though others could be cited.

First, the Common Ground Relief (CGR) is a community-initiated volunteer organization that provides short-term relief for victims, human and nonhuman, of hurricane disasters in the Gulf coast region, and long-term support in rebuilding the human and biotic communities affected in the New Orleans area by Katrina.[60] Originally formed in response to the immediate crisis created by Hurricane Katrina, the CGR is grounded in the value of cohabitation, offering assistance, mutual aid, and support to the residents (population, 14,000) of the Lower Ninth Ward of New Orleans, as well as to its coastal ecosystems. Devastation from the hurricane involved the loss of a community where more than 54 percent of the residents had lived for at least twenty-five years. Strong roots in the neighborhood resulted in the highest percentage of African American home ownership of any city in the United States.

The value of celebration, as well as the principles of recognition, participation, and empowerment, is evidenced in the efforts of the CGR in the day-to-day ways it, along with residents, struggle successfully to overcome the damage wrought to communities by Katrina. The CGR's volunteer-run projects range from construction and rebuilding to job training to advocacy and community outreach to community gardening and soil regeneration to wetlands restoration. One of the goals of the project is to re-create historical levels of agricultural self-sufficiency in the community through conducting soil testing, building raised garden beds for residents, organizing a local garden club, and providing gardening advice throughout the community.

The values of cohabitating, creating publicity, and accommodating variety are reflected in the ways in which the organization promotes a plurality of activities, in public places, that involve land and people from various walks of life. The Wetlands Restoration program, which restores native shore grasses and trees to area coastal lands, involves, for instance, a partnership with the local public school, a state university, the Louisiana Department of Agriculture and Forestry, and several local environmental organizations. Striving to rebuild the destroyed buffer of wetlands of coastal Louisiana and create awareness of its historical importance, the restoration efforts empower residents to participate in and care for their urban and coastal natural lands.[61]

Recently, CGR's wetland restoration program worked in partnership with the Restore the Earth Foundation and the Louisiana Department of Fisheries and

Wildlife to replant Gulf Coast wetlands in the wake of the 2010 British Petroleum Deepwater Horizon oil spill in the Gulf of Mexico.[62] This initiative involved deploying Gulf Saver Bags—biodegradable bags with a soil mixture containing oil-eating microbes, studded with Spartina marsh grasses and Black Mangrove—in the Pass-a-Loutre Wildlife Management Area, located at the southeast tip of Plaquemines Parish, where the Mississippi River delta empties into the Gulf of Mexico and provides a major fly-way for numerous migratory bird species.[63]

Exhibiting the value of cohabitation, the Gulf Savers work "in full cooperation with nature" to bring a "simple all-natural solution to accelerate the recovery process in the Gulf of Mexico's ecosystems."[64] Additionally, the Gulf Savers project exemplifies the values of creating publicity and accommodating variety in that it was implemented in public wetlands and wildlife areas and conducted entirely by (approximately thirty thousand) volunteers from a variety of organizations representing diverse populations. Volunteers construct Gulf Saver Bags, filling standard biodegradable burlap (sand) bags with an organic humus mix (rather than sand) into which native marsh plants are "plugged."

They then transport the bags by workboats, barges, airboats, and shallow draft boats to damaged wetland and wildlife areas, planting them in the muddy, washed-out coastal areas. Once Gulf Saver Bags are deposited in the wetland areas, the marsh plants' root systems begin to develop and break through the biodegradable bag. Within three months, the bags decompose, allowing the roots to expand into the existing sediment and hold together, protect, and restore the wetlands.

CGR is responsible for the recruitment of volunteers to assist in the construction and deployment of the Gulf Saver Bags.[65] Representing the value of sensuousness, the creation and implementation of the bags physically engages the restorationist in myriad sensory ways. Fingers touch burlap and shovel, hands scoop humus, arms lift plants, shoulders and legs hoist bags filled with soil and marsh grasses. The sun scorches Gulf restorationists' faces and arms, and mud sucks their feet and ankles. There are sounds of conversation, laughter, exhaustion, a shore bird, the Gulf wind and waves, and sights of the beauty, colors, contours, variation, and symmetry, as well as of the degradation and despoilment, of the seascape. The CGR's call for volunteer restorationists states the following instructions, again speaking to the quality of sensuousness: "You should expect to get muddy. . . . We require all volunteers to wear long pants, preferably not jeans as they become heavy and hot when wet. Volunteers in shorts will not be allowed to plant due to the presence of snakes and insects at our sites. . . . If any volunteers have rubber boots, we ask that they bring them if possible, although we will provide boots for those who do not have their own. Hats are also a good idea. . . . Make sure that everyone has access to sunscreen because the sun can be very intense no matter the season. Bug spray is also highly recommended."[66]

My second example of a restorative community of place is located in rural central Vermont. Vermont Family Forests (VFF) is a restorative forestry initiative that promotes the conservation of forest community health and the careful cultivation of local family forests for community residential benefits. Vermont's forests were severely depleted throughout the late-nineteenth and early-twentieth centuries because of intensive industrial logging practices, as was the case in many regions in the United States. Today, however, the forests have begun to recover, accounting now for nearly 80 percent of the state's land. About 70 percent of this forestland is privately owned—referred to by VFF as "family forests"—and used for nonindustrial purposes, placing individual landowners and communities in a pivotal role in the restoration and conservation of Vermont's natural lands.

Inspired by the work of conservation writers Wendell Berry and Aldo Leopold, VFF promotes the idea "that the three great conservers of family forests are well-informed forest stewards, sound economic returns from ecological forestry, and a community-shared land ethic."[67] Along these lines, it promotes restorative forestry practices that provide for human needs while preserving the forest's capacity to maintain itself as a healthy, natural ecosystem. VFF has created a program, for instance, that verifies and endorses firewood, flooring, and other forest products to help landowners receive sound economic returns for their ecological stewardship. Ultimately, it endeavors to inspire "an intense consciousness of land" and a community-shared land ethic through workshops, community celebrations, and publications.[68] To this end, education and community outreach are at the core of VFF's mission to empower landowners and residents in ways that help them gain the knowledge and resources necessary for healthy forest management.

Among the outreach work VFF emphasizes is helping landowners and townspeople connect experientially with their forestlands. Drawing on the value of *cohabitation*, the slogan "friend of the forest" has served as VFF's underlying mottos since its inception. Physically drawing residents into the ongoing management of locally owned forests, the VFF emphasizes the values of sensuousness and wildness of people's life places, as well as the principles of recognition and participation. VFF created a "Town Forest Health Check" as a way to strengthen connections between Vermonters and their forests and to encourage citizen-based forest stewardship.[69] The Town Forest Health Check provides an in-depth guide for nonspecialists to evaluate the current health of their forest and then, where needed, develop restorative ecoforestry practices in response.

The guide includes detailed descriptions of twelve benchmark forest health assessments, which include, for example, the evaluation of a forest's access paths and trails, stream crossings and conditions, tree species richness, nonnative exotic trees and shrubs, sensitive areas, small woody debris, legacy trees, snag and cavity trees, and large downed wood. There are photos illustrating key types of ecological degradation (e.g., soil erosion along access paths) and measurement techniques performed by landowners and community residents (e.g., two women measuring the diameter of a tree).

Forest stewards utilize a variety of low-tech tools (which VFF provides, and the guide describes with detailed usage instructions) to conduct the evaluation. "You don't need a lot of expensive equipment to get a good sense of what's going on in your forest," states the VFF.[70] VFF's Health Check Tool Kit includes a VFF cruising stick for measuring tree diameter, a protractor clinometer for measuring slope and identifying invasive plants, a penny for measuring the number of trees per acre, health tally check sheets for recording data, and a roll of bright flagging that serves as a second person when conducting measurements. Other helpful tools, states the guide, include number 2 pencils, permanent markers (for writing on flagging), a clipboard, a map of the property, a USGS 1:24000 topographic map, and the book *Wetland, Woodland, Wildland* by Elizabeth Thompson and Eric Sorenson.

Once the forest health assessment is completed, certain restoration activities are recommended by VFF. Activities, as with the tools just noted, are low tech but significant in the Vermont forests that VFF and local residents help restore and manage. Most critical is restoring and maintaining the health of soil in the forests, says VFF's founder and director David Brynn.[71] To this end, VFF spends a significant amount of time educating forest stewards about how to create erosion control structures such as waterbars, broad-based dips, and turn-ups on forwarding paths and skid trails used to transport logs.[72]

The installation of waterbars in particular, reports Brynn, is among the most basic and commonly used soil-restoration techniques in VFF forests. This involves measuring the grade of the trail and the downward slope (which needs to be 30 degrees or less, depending on the trail grade) in order to determine where the bar should be created. Installed by digging a narrow trench, waterbars are typically eight inches deep, with a two- to four-degree gradient (deeper waterbars, twenty-four to thirty inches, are created along roads that are closed to traffic). The outlet of the bar drains at a slight outslope away from the road and onto undisturbed vegetation or debris. Restorationists work from one end of the trail or road down to the landing (where logs are stacked) so that their completed work is not damaged.

Exemplifying the values of creating publicity and accommodating variety, as well as the principle of participation, VFF's forest health assessment process often involves community members from various backgrounds, ages, and interests. The guidebook is written and illustrated at a basic level, for example, so as to invite and encourage participation from nonspecialists. Spotlighting the importance of building a community-based land ethic, VFF organizes a Winter Solstice celebration marked by a community bonfire built in the middle of one of the forests it has worked to restore. Brynn also envisions the creation of green burial sites within family forests, where forest stewards can literally, physically become restored to and buried in the land that they have helped to restore and conserve.

These are but two examples of places that are enacting models of restorative community in ways resonant with the principles put forward here. I end this chapter with an opening. For despite the hopefulness that restorative initiatives such as these represent, the challenges and obstacles to developing restorative communities of place are equally present in today's culture. Novelist Wallace Stegner writes, "The challenge of forming cultures involving heterogeneous populations do not grow steadily from definable quality to definable quality.... They start from something, not from nothing. Habits and attitudes that have come to us embedded in our inherited culture, especially our inherited language, come incorporated in everything from nursery rhymes to laws and prayers, and they often have the durability of flint pebbles in puddingstone. No matter how completely their old matrix is dissolved, they remain intact, and we are deposited almost unchanged in the strata of the new culture."[73]

The same might be said in relation to the challenge of forming cultures with a restorative orientation. We have a hyperglobal, free-market economy to contend with, along with its homogenizing, materialistic orientation. Mobility, consumerism, and individualism reign in most communities of highly industrialized society, making it neither possible nor desirable, for various reasons, to live lives rooted in particular places—no matter how plural and open to difference and dissent those places may be. Developing restorative communities of place will need to be accompanied by the creation and narrating of new stories with accompanying individual and cultural-symbolic meaning related to a healing nature and humanity. Sketching part of this story and some of these meanings constitute the task of the following chapters.

NOTES

1. Higgs, *Nature by Design*, 185–95.
2. See Borgmann's *Technology*.
3. Higgs, *Nature by Design*, 255.
4. Ibid., 261.
5. Andrew Light, "Ecological Citizenship," 176.
6. Ibid., 177.
7. Ibid.
8. Ibid., 176.
9. Ibid.
10. Ibid.
11. Ibid., 182.

12. Ibid.

13. On this view of justice in relation to nature and humans, see, for example, Bullard, *Quest for Environmental Justice*; and Shrader-Frechette, *Environmental Justice*.

14. Various perspectives on the idea that nature has inherent value and thus ought to be treated with respect can be seen, for example, in Sylvan (Routley), "Is There a Need for a New, an Environmental Ethic?"; Singer, "Not for Humans Only"; Regan, "Animal Rights"; Paul W. Taylor, "The Ethics of Respect"; Rolston, "Value in Nature," all in Light and Rolston, eds., *Environmental Ethics*.

15. Implicit in Young's critique of contemporary communitarian thought is also a critique of some historical versions of communitarianism (e.g., Rousseau), though she does not focus her critique on ancient or medieval (e.g., Aristotle or Aquinas) versions of communitarian thought.

16. Young, *Justice and the Politics of Difference*, 12.

17. Ibid., 236.

18. Ibid., 230. Although Young rejects dominant communitarian proposals, she also, nonetheless, refutes the underlying social ontology of liberal theories of justice in terms of their individualist or atomistic conceptions of the self. This social ontology "presumes that the individual is ontologically prior to the social" and "usually goes together with a normative conception of the self as independent . . . autonomous, unified, free, and self-made, standing apart from history and affiliations, choosing its life plan entirely for itself." Young, as with liberal theories of justice more generally, upholds a universal value of the equal moral worth of all individuals as a basis of her view of social justice. Two basic values, both related to the principle of autonomy, define social justice for Young: (1) developing and exercising one's capacities and expressing one's experience, and (2) participating in determining one's action and the conditions of one's action. In this way, Young limits the work autonomy does in her account, preferring instead the notion of empowerment. Ibid., 45, 37.

19. See, for example, J. Baird Callicott, "Holistic Environmental Ethics," 111–25.

20. On this set of issues, see Bron Taylor, *The Encyclopedia of Religion and Nature*, s.v. "Ecofascism."

21. Marcus Hall, "American Nature, Italian Culture," quoted in Higgs, *Nature by Design*, 146.

22. Ibid.

23. J. O'Neill, "Time, Narrative and Environmental Politics," in Gottlieb, *Ecological Community*, 15; quoted in Higgs, *Nature by Design*, 146.

24. See the "SER International Primer," 3.

25. Exotic species are defined by the National Park Service as "those that occur in a given place as a result of direct or indirect, deliberate or accidental action by humans (not including deliberate reintroductions)." National Park Service, Management Policies (Washington, DC: U.S. Department of the Interior, 1988), quoted in Throop, "Eradicating the Aliens," 181.

26. It should be noted that invasive species are generally not a significant problem in the Vermont landscape, and most restorationists agree that invasive species

prohibiting the emergence or growth of indigenous or noninvasive nonindigenous species of the characteristic ecosystem should be removed to the greatest extent possible—on grounds that invasives disallow land from returning to its evolutionary historical character and trajectory. This said, the issue of nonindigenous, noninvasive species remains a controversial one within restoration ecological discussions. On the dilemma over whether invasive species are harmful to ecosystems, see the philosophical debate between Mark Sagoff and Daniel Simberloff: Sagoff, "Do Non-Native Species Threaten the Natural Environment?" 215–36; and Simberloff, "Non-Native Species Do Threaten the Natural Environment!" *Journal of Agricultural and Environmental Ethics* 18 (2005): 595–607.

27. Shore, "Controversy Erupts," *Restoration and Management Notes* 15, no. 1 (1997): 25–31.
28. See Mendelson, Aultz, and Mendelson, "Carving Up the Woods," 135–44.
29. Ibid.
30. See the "SER International Primer," 3, 8–10.
31. On this idea, see Jordan et al., "Foundations of Conduct."
32. Rosemary Radford Ruether, "The Biblical Vision," 1132.
33. Northcott, "Wilderness," 382–99.
34. Ibid.
35. Ibid.
36. Ibid.
37. Ibid., 397.
38. Ibid.
39. Ibid., 398.
40. Young, *Justice and the Politics of Difference*, 33. See also Charles Taylor, "The Politics of Recognition" in *Multiculturalism*, 25–73; and Walzer, *Spheres of Justice*.
41. According to Young, the normative ideal of city embodies four social virtues that represent heterogeneity rather than unity: social differentiation without exclusion, variety, eroticism, and publicity. The discussion that follows draws on, and goes beyond, several of Young's points related to these virtues.
42. See his *The Dream of the Earth*, 166–68.
43. Thayer, *LifePlace*, 3.
44. Berg and Dasmann, "Reinhabiting California," 217–18.
45. Fromm postulated five basic human needs: relatedness, transcendence, rootedness, sense of identity, and a frame of orientation. See his *Escape from Freedom*.
46. See Beatley, *Native to Nowhere*.
47. The following discussion on communities of place draws on Young's and Beatley's conceptualizations of ideal cities and place-oriented communities, respectively.
48. Low and Altman, "Place Attachment," 5, quoted in Beatley, *Native to Nowhere*, 26.
49. Ibid., 28.
50. Farley, "Religious Meanings," 110.
51. Young, *Justice and the Politics of Difference*, 238–39.
52. House, *Totem Salmon*, 190.

53. Jordan, *The Sunflower Forest*, 89–90.

54. For a fuller description of these dimensions of the human connection to nature, see Kellert, *Kinship to Mastery and Birthright*.

55. See, for example, Beatley, *Native to Nowhere*; and Thayer, *LifePlace*.

56. Mills, *In Service of the Wild*.

57. Young, *Justice and the Politics of Difference*, 240.

58. Ibid.

59. Thomas Berry, *Dream of the Earth*, 168.

60. For the description that follows, see the CGR website, www.commonground relief.org.

61. Common Ground Relief, "Our Projects," www.commongroundrelief.org/node/2 (accessed June 20, 2010).

62. See Common Ground Relief, "Wetlands Restoration," www.commonground relief.org/gulfsaverinitiative, and Restore the Earth Foundation, www.restorethe earth.org/ (accessed April 24, 2012).

63. This project implemented the first major restorative wetland plantings to be permitted by the US Army Corps of Engineers (USACE) following the Deepwater Horizon oil spill. See Restore the Earth Foundation, www.restoretheearth.org/ gulf_savers.html (accessed April 24, 2012).

64. Ibid.

65. CGR also expanded its community nursery in order to grow plant material for the bags, as well as native trees for planting throughout the region.

66. Common Ground Relief, "Wetlands Restoration Volunteer Info," www .commongroundrelief.org/wetlandsvolunteerinfo (accessed April 24, 2012).

67. On Berry and Leopold, VFF states: "Wendell Berry once wrote that 'the two great ruiners of private land are ignorance and economic constraint.' And Aldo Leopold wrote that 'perhaps the most serious obstacle impeding the evolution of a land ethic is the fact that our educational and economic system is headed away from, rather than toward, an intense consciousness of land.'" Vermont Family Forests, "About Us," www.familyforests.org/about/ (accessed June 20, 2010).

68. Vermont Family Forests, "Family Ecoforestry," www.familyforests.org/ecoforestry/ index.shtml (accessed April 24, 2012).

69. See "Town Forest Health Check: A Town Forest Steward's Guide to Forest Health Assessment," www.familyforests.org/ecoforestry/TFHC/TFHC_Guidebook_ web.pdf (accessed April 24, 2012).

70. Ibid.

71. Personal communication, April 2008.

72. For a description of these restoration techniques, including the one below on waterbars, see Natural Resources Conservation Service, "Forest Trails and Landings: Vermont Conservation Practice Job Sheet 655, October 2011," available at http://efotg.sc.egov.usda.gov//references/public/VT/JS655VT_FillableForm .pdf (accessed April 25, 2012).

73. Stegner, *The American West as Living Space*, quoted in House, *Totem Salmon*, 190.

PART II

Restored To Earth

CHAPTER FIVE

༄

ECOLOGICAL SYMBOLIC ACTION

Restoration as Sacramental Practice

*What restoration could and should be for us is the trans-
formation of our souls. In addition to what this work may
accomplish in the land, I yearn for it as the yoga that will
cause us to evolve spiritually, that will restore to us a feeling
of awe in something besides our own conceits.*

—Stephanie Mills, *In Service of the Wild: Restoring and
Reinhabiting Damaged Land*

In the previous chapter we explored some of the ways in which ecological restoration practice could serve as a context for developing vernacular communal
values in relation to particular places. Further, we saw how some communities use
restoration activities as a way to strengthen social connections between people
and ecological connections with natural systems. In this chapter I examine how
restoration functions as a form of symbolic action, as well as a type of social and
ecological practice, in relation to healing damaged natural lands. Restoration may,
in other words, attempt to regenerate healthier ecosystem processes and place-
oriented communities, but it may also come to mean much more than this—for
participants and for broader society. As the title of this book suggests, through
restoring earth, people and communities may become *restored to earth*.

The idea that restoration activities may serve both scientific and symbolic
functions is not without controversy within restoration thought and practice. On
the one hand, the view that a scientific ecologically based practice such as restoration
may also involve a meaning-making function raises fundamental questions regarding the ways in which science, religion, and ethics are and should be connected.
On the other hand, some restorationists, secular and religious, utilize explicitly

religious-oriented language, albeit in varying ways, to describe the experiences and broader cultural implications of restoration.

One of the purposes of this chapter is to explore and describe the paradoxical ways in which symbolic action functions in ecological restoration thought and practice. That is, the kind of action inherent to ecological restoration practice involves fundamental experiences and ideas that, while seemingly contradictory, may also provide deeper insight or truth regarding meanings for nature and the human relationship to nature. The paradoxical themes of death and life, the sacred and the profane, and discovering and constructing nature, for example, run through the restoration enterprise and thinking.

In addition to this descriptive task, however, this chapter also develops a normative proposal. I provide a potential model for restorative symbolic action—one that is instructive for all ecological symbolic action in some sense. Previous chapters have examined the meanings for nature (chapter 2), direct personal experiences (chapter 3), and communal values (chapter 4) that the practice of restoration may yield. In this chapter I lay a theoretical foundation for the constructive vision that follows in the next chapter, arguing that a symbolic interpretation of concrete ecological acts such as restoration, and the direct experiences they may yield, are integral to the formation of cultural environmental values—and to the creation of a broader restoration narrative and story (the focus of chapter 6).

I begin this chapter by critically analyzing some of the ways in which restoration and religion scholars have variously understood the concept of symbolic or ritual action in relation to nature. Next, I propose a model for interpreting restoration as a form of symbolic action based on the restoration work and experience of a practicing religious community. Finally, I suggest two ways in which the activities of restoration can be further understood as performing particular types of symbolic action that create broader cultural meanings in relation to the human acts of healing damaged land.

ECOLOGICAL SYMBOLIC ACTION

The idea that restoration can function as a transformative activity as Mills's text at the outset of this chapter suggests requires that people are, in fact, participating in restoration efforts. Public, volunteer participation in restoration projects, however, is not always present. As noted in my field guide to types of restoration (chapter 1), much restoration activity today requires a certain amount of scientific and technical knowledge as well as skilled engineering and design input, which means that restoration is often performed by specialized firms or government agencies with a cadre of professional scientists and conservation practitioners.[1] Ecological restoration today, as we have noted, runs the risk of technological drift, that is, of becoming a predominantly scientific–technical practice dominated by experts and technicians.[2]

Despite the technological drift in restoration activities, many restoration projects today, including most of the projects referenced in this book, use considerable numbers of volunteers. For instance, the Midewin National Tallgrass Prairie project in Joliet, Illinois, the former site of the Joliet Army Ammunition Plant, which required extensive cleanup from contamination from decades of TNT manufacturing and packaging, relies heavily on volunteers to maintain more than fifteen thousand acres of tall grass prairie. And the restoration projects of Common Ground Relief in New Orleans, Vermont Family Forests in Bristol, and Holy Wisdom Monastery in Madison, to name a few, rely extensively on volunteers from the community to restore coastal wetlands, forests, and prairie and oak savannah ecosystems. To counter overly technological approaches to restoration, some authors, including Higgs, Light, and myself, have proposed that classifying ecological restoration as good or excellent must include consideration of values besides scientific and technical ones.[3] Recall Light's argument from the previous chapter, for instance, that good restoration requires the generation of moral value along with natural value in the implementation of a particular restoration project. In other words, in addition to creating ecological integrity (natural value), a good restoration project will also promote ways for individuals to reconnect with nature (preferably local), namely through public participation in restoration projects (social value).

This is not to say that professionals and experts are unnecessary in restoration projects, as we have seen. In fact, professionals are almost always involved in some way in restoration projects; the mix of scientists, practitioners, activists, and volunteers create one of restoration's unique aspects as a contemporary environmental practice. Nevertheless, as I have argued in the previous pages, there are social, communal, spiritual, and moral values that restoration practice can form— for instance, connecting individuals with nature through hands-on experiences, promoting a sense of care for particular landed places—that can be accomplished only through a certain level of volunteer, public participation in a project.

With this public participatory type of restoration in mind, I would like to examine more concretely some of the ways in which restoration may function as symbolic or ritual action in relation to land, an aspect neither Higgs nor Light considers in his framework. There are at least two characteristics of restoration practice that may contribute to its potential to create broader symbolic meaning in relation to nature: the actual practical work of restoring damaged land and the larger symbolic work that such activities perform.[4] Not only are the daily (or weekly or seasonal) activities of restoration work "just the work that needs to get done" as one Vermont restorationist told me, they also often signify and mean much more than that to restorationists themselves and to the broader community and society. In other words, the spiritual and moral experience and broader cultural and religious meanings that restoration practice enables begin in the practical ethical action of restoring nature itself. It is through the actual hands-on activities of restoring land—ripping out, replanting, reintroducing, rebuilding, reactivating, and so on—that people experience restoration's transformative potential. "Acts transform people, and this act [of restoration] transforms people in a particular way," states William Jordan.[5]

The idea that acts transform people is, of course, not a new idea. Aristotle and, later, Thomas Aquinas wrote about the concept in relation to their virtue theories and, in particular, in relation to how moral virtue is formed within people through the habituated performance of this or that virtuous act. Similarly, restoration practice involves activities, such as careful, patient, long-term observation of ecosystems, that may help individuals become attentive to land's distinct history, features, and functions. Chicago Wilderness restoration volunteers, for instance, will often send email messages with long lists of native and invasive plants in their areas, all identified by Latin scientific name. Further, practicing restoration activities may help participants acquire a deeper understanding of how much more time consuming, costly, and laborious it is to repair ecosystems than it is to degrade them; it may also help form a renewed sense of confidence and trust in the self-healing capacities of land when it is given sufficient time to recover.

Even as restoration practice may yield direct experiences and ecological knowledge in relation to particular ecosystems, it may also create larger symbolic meaning, perhaps even religious meaning in relation to nature and its healing. Jordan argues, as already noted, that restoration may be able to provide the context for a new ritual tradition in relation to our natural landscapes.[6] The ritual tradition Jordan envisions will provide a contemporary context for humans to enter into deep community with Other Nature.[7]

Environmentalism since Emerson and Thoreau has privileged unity with nature over difference, oneness over otherness, and harmony over ambiguity, argues Jordan, as noted previously.[8] Jordan suggests that this bias has precluded modern humans from entering into the deepest form of community with natural ecosystems. "Nature may be beautiful. It may even include an element of altruism," Jordan states. "But it is also terrible and shameful, and it is only by confronting and coming to terms with this—what the mythologist Joseph Campbell called . . . 'the monstrosity of the just-so'—that we achieve the deepest kinds of relationship with it."[9] Along compatible lines, religious environmental ethicist Lisa Sideris argues that ecological theologians have privileged overly positive notions of evolutionary theory (cooperation, community, perfect adaptation) over and against more negative ones (struggle, suffering, death, disease). Sideris argues that the oversight of evolutionary theory's more negative aspects has led to less than satisfactory, romantic, ecological theological models.

Jordan outlines four steps that characterize the move into "real" community (adapting Emerson's and Thoreau's romantic view of community) between humans and land. The first step involves gaining awareness of the others in our midst, the "infant awakening" that we live as inhabitants in a world populated by human and nonhuman others. Second comes interaction and participation with these others through active engagement in the ecologies of particular natural landscapes. This active participation in ecological processes, beyond simply observing or recreating in nature, becomes the necessary foundation for the cycle of gift exchange, or mutual reciprocity, that is integral to any deep relationship. The third step into genuine community, according to Jordan, involves the gift humans may give back to land, namely, the restored or reactivated ecosystem. This, writes Jordan, "is perhaps as close as we can come to paying nature back in kind for what we have taken from it" and continue to take from it.[10] Finally, as in all gifts given in relationships, human–human and human–nature, the act of restoration is never entirely adequate; that is, true ontological unity is never ultimately possible, according to Jordan, given that the notion of community itself involves a multiplicity of subjects, as well as a distinction at some level between self and other.

Difference, not unity, for Jordan, lies at the heart of communion. This idea resonates with the work of feminist ethicists, as we saw in the previous chapter, that critique dominant theories of community for their overly positive assessments of community life, to the neglect of its more negative aspects. "Communities harbor as much potential for harm as for hope," writes theologian Serene Jones, and feminists "usually begin their assessment of a community by looking at its fractures and its potential for violence."[11] Further, dominant interpretations of community frequently overlook the diversity of voices and moments of conflict and resistance within community life. "By listening to a diversity of voices, feminists argue, one better appreciates the ongoing conflicts that marking living traditions and communities."[12]

With regard to the act of restoration, conceiving it as a ritual performance—firing a prairie, for example, though with some sort of ceremony attached—makes the act of gift exchange between humans and Other nature not perfectly harmonious, but more reflexive, enhancing its capacity for generating the deepest forms of values (as was the case with firing the prairie on Maundy Thursday at my children's elementary school in central Illinois, noted in chapter 3).[13] Jordan proposes that ecological restoration can be seen as a potential new ritual tradition analogous to the world-renewal tradition of, for instance, the Australian aboriginal people and their annual ritual of singing the world back into being. Additionally, Jordan, along with others, have viewed performance and ritual (e.g., artwork, festival, celebration) incorporated within restoration projects themselves as positively contributing to restoration's capacity for meaning making and community-building. "At the deepest level," writes Jordan, "ritual offers the only means we have of transcending, criticizing, or revising a morality or ethical formulation prescribed by authority or handed down by tradition. Most fundamentally, it is the means by which humans generate, recreate, and renew transcendent values such as community, meaning, beauty, love, and the sacred, on which both ethics and morality depend."[14]

The ritual dimension of ecological restoration is especially important for environmentalism, Jordan proposes, because it suggests that reflective action and concrete place-based experience serve as a basis for the creation of meaning and environmental values, rather than deductive, top-down approaches.[15] Jordan cites "Mass on the world," by French philosopher and Catholic priest Pierre Teilhard de Chardin, as an example of earth-based ritual action he views as necessary for the creation of higher human values in relation to land.[16] "If the Mass in its traditional form is rooted in the act of sacrifice as a way of dealing with the horror of creative death, figured in the killing of a single creature, or even of God," writes Jordan, "then a Mass on the world would provide a way of dealing, in symbolic terms, with both the killing and the resurrection of entire ecosystems." Restoration should be conceived as a type of public liturgy, Jordan proposes. It "must become a community event in a way that backpacking cannot."[17]

Yet one wonders how productive a symbol a "Mass on the world" is in contemporary evolutionary ecological terms. It may be, as Teilhard believed, that the death and resurrection of Jesus Christ involves all matter—real death, but also real transformation from death to life in an evolutionary sense. As he writes, the "consecration of the world would have remained incomplete, a moment ago, had you [God] not with special love vitalized for those who believe, not only the life-bringing forces, but also those which bring death."[18]

Still, the sacrificial, redemptive act of Christ has traditionally emphasized a supernatural *overcoming* of earthly, biological death, rather than its brute reality in evolutionary ecological terms. Teilhard's type of symbolic interpretation of creation may adequately point to the paradoxical aspect of biological life itself, with its perpetual pattern of transvaluing death into life and life into death.[19] Yet even as it does this—that is, seriously considers the realities and necessity of death in the evolutionary system—it nevertheless views these brute realities as negative, something to be ultimately overcome by divine love. "The man who is filled with an impassioned love for Jesus hidden in the forces which bring death to the earth," states Teilhard, for instance, "him the earth will clasp in the immensity of her arms as her strength fails, and with her he will awaken in the bosom of God."[20]

The inherent transvaluation that occurs in evolutionary, ecological, historical life involves necessarily pain, struggle, suffering, and death—not as something to be essentially, supernaturally overcome, but as a quality that is intrinsic to, unavoidable in biological life. It is what is: the beauty and tragedy of prolific, generative, wonderful life, as Holmes Rolston is fond of stating. Death is really death in the ecological evolutionary system. There is natural pain and suffering that have to be.

I want to be careful here in juxtapositioning the "suffering that has to be" in evolutionary biological terms (e.g., death) and the "suffering that does not have to be" in social cultural terms (e.g., racism). Religious environmental ethicist Larry Rasmussen cites this distinction in his "Returning to Our Senses: The Theology of the Cross as a Theology for Eco-Justice."[21] Yet Rasmussen focuses on the latter type of suffering in interpreting Christ's sacrificial action on the cross, neglecting to consider how the darker side of evolutionary ecological theory shapes theological notions of suffering—and of redemption.

The ethical question with regard to a symbolic ecological interpretation of the natural world becomes: How does one move from an overly pragmatic acceptance of the character of evolutionary biological life, to a deeper, more reflective, perhaps even sober celebratory recognition of its wondrous, miraculous function? Ecological theologian H. Paul Santmire begins to provide a Christian response to the issue of "ritualizing nature."[22] Though it is not so much the ritualizing of nature itself that is the focus of a restoration ethic. Rather, it is the ritualizing of the active, and at times ambivalent, encounter between the human self and the other members of the land community that is the focus of concern. Restoration accentuates the reality that learning to live more graciously and cooperatively with land is an ambiguous act of struggle and conflict as much as it is a joyful act of celebration and freedom. Along these lines, a religious restoration ethic asks: What is the value, or even virtue likely to be achieved in the human participation in the fundamental acts of creation, planting, harvesting, killing, burning, for instance, as a way to reverse past human acts of de-creation?

Jordan's notion of restoration as a new ritual tradition can be helpfully illuminated, and taken in a new direction, by expanding on Mills's reference above to the spiritual practice of yoga in relation to restoration. Similar to modern yoga, the spiritual practice of restoration can be understood as secular, or public, in nature. Yoga scholar Elizabeth de Michelis, for instance, proposes modern (postural) yoga as a "healing ritual of secular religion," one that can offer "some solace, physical, psychological or spiritual, in a world where solace and reassurance are sometimes elusive."[23] This healing ritual, and the solace it may yield, depends not on one's existing (spiritual, religious, or secular) beliefs and commitments but rather, on the habituated act of the practice itself. Restoration too can be understood along these lines. Nevertheless, given its explicitly ecological basis, as well as its enactment out in the open—in the fields, forests, woodlots, and wetlands of society—it is more apt to call restoration a healing ritual of public ecology; or, as I propose calling it, a public ecological spiritual practice.[24]

Additionally, yoga and ecological restoration serve as secular or public spiritual practices in terms of their capacity for creating meaning in individuals' lives, meaning that is formed in and through the practice's actual, embodied activities themselves. In relation to yoga, Joseph Atler writes, for example, that "the possibility of transcendence is dependent on Life itself, as Life is experienced through the body by a person who practices Yoga."[25] In restoration's case people are drawn into and experience nature's relentless life force and capacity for regeneration. Restorationists find meaning, a home, within particular ecosystems and their slow, self-healing ways. Further, yoga and restoration importantly form meaning and a sense of belonging through the development of "shared communities of practice."[26] Where practitioners interpret the work in spiritual terms—for "there is room for the practitioner to decide whether to experience her practice as 'spiritual' or as altogether secular," as de Michelis points out—restoration, as well as yoga, can be seen not only as a solo or private spiritual practice but also as "shared communities of spiritual practice."[27]

Understanding ecological restoration as a type of public spiritual practice that has the capacity for making meaning in relation to particular places and communities of people resonates with Benedictine restorationists' understanding of restoration as a type of ritual practice. Despite the fact that these restorationists interpret restoration activities in explicitly religious terms, their focus on restoration as an embodied, communal, meaning-making activity available to all who engage in its hands-on practices exemplifies a form of what I have just defined as public spiritual practice. Furthermore, the community-based orientation of their restorative work draws volunteers from various backgrounds and beliefs, creating the values of restorative communities of place (e.g., cohabitation, publicity, and variety) that we explored in the previous chapter. It is thus here that we may find a potential model for restorative symbolic action, one that may, in some sense, be instructive for restorative symbolic action more generally.

RESTORATION SYMBOLIC ACTION AS SACRAMENTAL PRACTICE

The Benedictine sisters at the Holy Wisdom Monastery in Madison, Wisconsin, have been working restoratively with the natural lands of the monastery since the community set down roots on a hill overlooking Lake Mendota in 1953. The monastery's original 40 acres consisted of farmland cleared and plowed in the early 1900s. Today, Holy Wisdom Monastery includes 130 acres with a 10,000-year-old glacial lake, wooded nature trails, restored prairie, gardens, and orchards.[28]

The Benedictine restorationists refer to land restoration as an activity that flows from their sense of sacramental spirituality and practice. Recall that the sacramental tradition of Christianity emphasizes the notion that the divine is in some sense present and discernible within the material, physical world. Restoration activities, according to the nuns, can awaken individuals to the nonhuman beings in their midst as well as to the divine presence within creation. The Rule of St. Benedict, which Benedictine religious communities follow, emphasizes the importance of daily practices in helping members of the religious community become more aware of God's presence in ordinary activities and in the larger world.[29]

The "balance of the day" is a central theme in Benedictine spirituality. In Benedictine monasteries such as Holy Wisdom, the day is structured according to periods, or "hours," with time allotted for work, prayer, study, and leisure. Restoration activities at the monastery can fall into each of the periods of the day according to the Holy Wisdom prioress, Sister Mary David Walgenbach.[30] For instance, "working in the prairie is difficult labor," says Sister Mary David.[31] "I view it as work."

The sisters, along with volunteers from the broader community, work to remove invasive species and collect and plant seeds in the monastery's restored prairie. "Weeding, pruning, digging are time-consuming, never-ending activities," says one volunteer at the monastery.[32] Seed collection, for instance, is a labor-intensive activity that requires walking in the fields for hours at a time in order to hand-strip or trim seeds from native prairie plants, one at a time. This activity usually occurs in the fall, when seeds are ripe and plants are wilted and dry. As with picking berries, restorationists typically wear buckets tied around their waist, into which they place the seeds and their casings. After the seeds are collected, they must be sorted (by species) and sifted or winnowed from stems, plant material, chaff, or broken seeds. They are then dried on racks, which are, over the course of a couple of months, periodically brought out into the sun so they will not mold. The seeds can then be used for replanting particular areas of the prairie.

Even though the prairie serves as a context for engaging in restoration work for some, at the same time, Sister Mary David says, others go to the prairie "to study the birds and plants, to learn their names and observe their ways."[33] In this way Holy Wisdom Monastery is a living laboratory for area students and environmentalists, who benefit from the bountiful prairie and the monastery's knowledgeable environmental coworkers. Teachers gain valuable hands-on environmental training to supplement their classroom education, while students of all grade levels frequently come to the monastery to learn about natural lands. Holy Wisdom's plans include the development of an outdoor teaching area near the restored prairie, oak savannah, and wetland preserve.[34]

In addition, Sister Mary David finds that many people go to the prairie for contemplative prayer, viewing it as "a time to renew and lighten their inner spirits."[35] Nature trails meander through the monastery's regenerated prairie and oak savannah, and along the restored glacial lake, as a way to foster contemplative experience in the natural world. One trail description, for instance, reads: "As you descend this trail, a glacial kettle lake hidden beyond the hillside appears. It attracts abundant wildlife and offers you a place for solitude and reflection."[36] Former director of the Leopold-founded University of Wisconsin at Madison Arboretum, Greg Armstrong, writes poetic "Nature Notes," posted on the monastery's website, which often express the contemplative benefit of the monastery's nature trails and restored lands.[37] On this, he writes:

> Seeds are falling all over the place, during this time of year. They are little baby plants with a lunch packed to sustain them over their time of rest and at the time of germination, until their leaves take over the task of supplying food. A lot of seeds do not fall straight down under their mother, where they would have to compete with her in the future, but have devised strategies through evolution for distributing their babies far and wide to increase the chances for success. Some of the seeds are produced in tasty fruit that are eaten by animals of various sorts and thereby carried and deposited far away. Some seeds have developed helicopter-like wings such as maples. Others have wonderfully clever parachutes such as the dandelion, which carry the seeds on the wind. The effectiveness, sophistication and diversity of these strategies for survival are nothing short of divine. . . . Come out to Holy Wisdom Monastery to visit and observe the miracle of life this fall! Walk the nature trails or sit on the benches overlooking the prairie or Lost Lake and feel the peace surround you.[38]

The monastery holds seasonal workdays where volunteers from the surrounding community can share in the restoration work of the monastery's natural lands. The morning of the workday is marked by a nature-related program such as bird watching or plant identification. This is followed by shared prayer focusing on the interfaith theme of the beauty of the earth. Then the group of one hundred or so volunteers shares in the restoration work that needs to be done on the ground's 130 acres. Environmental volunteers may work in the prairie to collect seed or remove invasive species, or they may pick apples, split logs, prune bushes, or maintain nature trails. The workday finishes with lunch at the monastery. It is a chance, say the sisters, "to come together with others at Holy Wisdom Monastery and enjoy the community spirit while tending to the earth."[39]

"The Benedictine community animates this place," Sister Mary David told me.[40] "Restoration work is a form of community building and inviting people in."[41] In this spirit, Holy Wisdom also holds an annual benefit concert for the land called Prairie Rhapsody. Local businesses and individuals sponsor the event, which brings well-known musicians to the monastery for an evening focused on "Saving the Land and Hearing Great Music," as one music reviewer called it.[42]

"From the hearty applause that came between movements," wrote the reviewer, "you could tell that two crowds were present: what I will call 'The Music People' and 'The Land People,' including of course many crossovers. But what brought both of them together were two things: the pleasure of great music wonderfully explained and performed; and the cause of restoring the land."[43]

Learning, praying, working, sharing at the table, and celebrating community form the framework for restorative symbolic action for the Benedictine restorationists. Moreover, this model of restorative symbolic action involves the idea that all willing and able individuals can work to restore—and find meaning through participating in—nature's abundance. The nuns at Holy Wisdom Monastery may explicitly interpret restorative activities as spiritual work, though people of all persuasions are welcome to the table. The work exemplifies public spiritual ecological practice, blending religious and secular sensibilities in a context where the whole biota is viewed as the moral community.

If one adopts the view that the environmental crisis includes fundamentally a crisis of the human spirit, character, or culture, as does this book, shared ways will need to be fostered in which people from various religious and political perspectives may become more integrally connected to their natural ecologies for their survival and fulfillment. That restoration practice may provide such a context further contributes to its distinctiveness as a contemporary environmental practice. A religious environmental ethic will need to consider explicitly the ways in which the act of restoration may form types of broader cultural symbolic values in relation to natural landscapes. This is a subject we examine in detail in the following chapter. For now, however, we turn to examine two ways in which restorative symbolic action might be further interpreted in relation to the concrete practice of healing damaged land.

TOUCHING THE HEART OF CREATION: RESTORATION AS SACRED WORK

Ritual studies scholar Catherine Bell proposes the term "ritualization" as "a way of acting that is designed and orchestrated to privilege what is being done in comparison to other, more quotidian activities.... [It is] a matter of variously culturally specific strategies for setting some activities off from others, for creating and privileging a qualitative distinction between the 'sacred' and the 'profane.'"[44] In the practice of restoration, it is the case of setting off and privileging the acts of restoration as qualitatively meaningful activities in themselves, in contrast to, say, sitting and playing video games on a Saturday afternoon or going shopping at the mall. While one can scarcely get more mundane than clearing brush, sorting seeds, planting trees, weeding, collecting water samples, these activities are raised to the level of the meaningful work of restoration when they are ritualized.[45]

But what is it more exactly that makes such acts meaningful, perhaps even sacred in restoration's case? What is it more precisely about restoration activities that make restorationists in the Volunteer Stewardship Network of the Chicago Wilderness Project, for instance, say that restoration work is "right up there" with raising children and making art, "more important to a lot of people than their jobs," as Laurel Ross, the network's coordinator says?[46] Why is it that, in a study of 306 volunteer restorationists in the Chicago Wilderness project, participants felt that restoration offered them meaningful action, creating deep satisfaction and a sense of personal growth?[47] Initially, one might say that restoration activities are thought of as meaningful because they are needed for repairing damaged ecosystems; they are meaningful, in other words, because the restorationist is working to promote an environmental value, that is, ecological health, for the good of people and land. Yet, there are additional reasons that restoration activities can be viewed as meaningful, even sacred.

In the first place, restorative acts are thought of as meaningful because they put individuals in close connection with nature—and, more specifically, close connection with nature in a particular way, that is, a participatory, active way. In the second place, they are thought of as meaningful because they put us in close contact with other individuals—and, more specifically, with other people performing the particular ethical action of restoring nature. Recall Vermont restorationist Marty Illick's comment, "It is the most uplifting thing. We are really just high being out there in nature working with a small group of people," as well as other restoration projects I have highlighted—my children's elementary school in central Illinois, Common Ground Relief in New Orleans, Holy Wisdom monastery in Madison.[48] In each of these, as well as community-based restoration more broadly, sociality is important to this nature-focused practice. This aspect casts new light on American conservation thought, for despite Leopold's primary interest both in people's relationship to land and in people's relationship to one another, it has tended to focus on nature and the human relationship to nature, but not on intra-human relations in land-based activities.

Further, for some religious restorationists, restorative acts are viewed as sacred activities because they provide a way in which individuals can faithfully follow or commune with the divine presence within nature. The restorationists of the Holy Wisdom Monastery, for example, state that "the divine is present in all creation" and that restoration work can help people "connect with the dynamic beauty of nature."[49] "Care for the earth comes out of our spirituality," they state. "People are inspired by their experiences here."[50] "They discover a spiritual connection with the land. Whether caring for it as an environmental volunteer, walking the trails, or spending time on a bench reflecting on the mysteries around them, there is something wonderful and amazing about having a place preserved as nature intended."[51]

Deep ecological–bioregional restorationists too view nature as holding intrinsic, sacred value, with restoration providing a context for people to commune with earth's sacredness. According to deep restorationists, nature and its beings, including humans, are sacred by virtue of their inherent evolutionary capacity for self-renewal, development, and self-realization; for Benedictine restorationists, nature is sacred by virtue of its capacity to connect humans with God's presence. As Sister Mary David states, "creation has a way of waking us up to the divine—of things we didn't create."[52] Restoration, in attempting to regenerate and give back to nature its natural historic wildness is "a profound obeisance to Nature," as Stephanie Mills writes.[53] For humans it is a "wide opportunity for enjoyment in the land, a sense of serving the sacredness of Nature, and touching it with your hands."[54]

Although all of earth and its beings have sacred value for deep restorationists, it is particular life-places, most often conceptualized at the bioregional level, and the species and beings that inhabit them, that warrant utmost respect and restorative care. Similarly, for the Benedictine sisters, the life-place of the monastery and its surrounding ecosystems take on sacred meaning.[55] Following St. Benedict's emphasis on the principle of stability, the Benedictine nuns "put down roots, for a lifetime," says Sister Mary David.[56] "We get to know the land, the creatures, plants, sky, and sun where we reside. . . . Our ecological sensitivity grows out of the Benedictine principle of stability and the commitment to live a sacramental life."[57]

Bioregional and Benedictine restorationists are at times eloquent narrators of the lived realities of species native to their particular life-places and what it means concretely for humans to learn from other species how to become truly participating residents within the land community. For example, Freeman House, as we have seen, writes poetically and passionately about the "life lessons from another species" he and a small group of committed residents of the Mattole Valley in northern California have learned over the past decade through working to restore the native Chinook salmon to their home river, the Mattole.[58] House recalls a salmon encounter he had early on in the river-restoration process. It was the night of New Year's Eve, 1982, and House was alone in a sixteen-foot trailer along the river, near the weir (the gate with holding pen inserted at the mouth of the river) they had constructed to trap incoming salmon so they could then harvest, fertilize, and incubate eggs, reintroducing the wild fingerlings back into the stream.

The Mattole Valley restorationists had constructed various styles of weirs before they finally developed a design that they felt served "the ends of the other species," rather than those of human comfort.[59] Although the pen of this new weir was more spacious and comfortable for the fish, it was nevertheless less secure. This meant that the restorationist on call in the trailer, House on this night, needed to check it more frequently to successfully capture a female salmon. This required setting an alarm clock that would wake the restorationist attendant at two-hour intervals, no matter the time of day or night.

This particular New Year's night it was midnight and raining when the alarm clock woke House. He pulled on chest waders over his long johns and jeans, and then a raincoat, a headlamp, and a black wool cap. Where the trailer was located a few miles upstream from the Mattole's headwaters, the river was more like a large creek with steep banks. The fish were wary at this late, spawning part of their lives, and so they often ran at night, in murky water, skittering quickly into deep holes or under logs at the slightest sound or sight of potential danger.

Painfully careful not to make any noise, House inched "down the bank crab-wise in wet darkness, the gumboot heels of the waders digging furrows in the mud, the fingers and heels of [his] hands plowing the soaked wet duff." Once at the bottom of the ravine along the bank of the river, he held his breath to let his ears readjust to the sounds of the water, straining to perceive any sound that might signal that a fish was trying to enter or was in the weir's pen. He heard a "plop, plop, plop," which is often the sound of a fish searching against the pen for an opening upstream. He held his breath again and waited. He heard the noise again and reached for the parachute cord that held up the gate of the pen of the weir. He yanked the cord, the gate dropped. About what ensued next, House writes:

> When I hear the sound I am waiting for, it is unmistakable: the sound of a full-grown salmon leaping wholly out of the water and twisting back into it. My straining senses slow down the sound so that each of its parts can be heard separately. A hiss, barely perceptible, as the fish muscles itself right out of its living medium; a silence like a dozen monks pausing too long between the strophes of a chant as the creature arcs through the dangerous air.... On this mindblown midnight in the Mattole I could be any human at any time during the last few millennia, stunned by the lavish design of nature. The knowledge of the continuous presence of salmon in this river allows me to know myself for a moment as an expression of the continuity of human residence in this valley. Gone for a moment is my uncomfortable identity as part of a recently arrived race of invaders with doubtful title to the land; this encounter is one between species, human and salmonoid.[60]

Restorationists such as House emphasize the loss of sacred wildness and eco-logical connection that occurs when life-places are damaged. House, for instance, writes of "ecosystem absences," the "ghosts of lost creatures," "globes of emptiness floating through the bloodstream of life" that occur when there is an absent life form within the web of life.[61] These ghosts "can become a palpable presence, a weird stillness moving against the winds of existence and leaving a waveform of perturbation behind."[62] Ecosystems and organisms are literally, palpably desacral-ized through human degradation of land according to religious restorationists.

The quality of given, sacred wildness has been diminished. Earth's sacred, interconnected web of life has been wounded.

Similarly, Sister Mary David recalls a situation where a golf course developer offered to buy some of the monastery's natural lands. The developer was interested in purchasing a particular hilltop area that had been restored within the grounds of the monastery. "That's God's spot," the nuns told the developer. "Even though we are only twenty minutes from Madison, we haven't sold out," Sister Mary David told me, implying that developing the hill or the monastery's restored natural lands would degrade spots where God's presence was known in an especially poignant way.[63]

Even as religious restorationists often mourn ecosystem absences, they also soberly celebrate and trust the seeds of sacred wild life and the potential for self-healing within land that remains, even where species have been lost. Nature and its wild ways are in a sense irrepressible, beyond the human capacity for domination and mastery. For example, even where native salmon, in House's case, were virtually wiped out, there remained seeds of generative life; feeder streams were clogged with silt and debris from intensive logging, as well as river banks eroded from lack of healthy vegetation, yet wildness remained—and could still be regenerated. There were plants and grasses, animals and birds, wildflowers and forests; there was still energy with nutrients flowing through the system and genetic information looping back and around. Wild, prolific life was still reproducing, growing, dying, and reemerging.

Where ecosystems and their organisms have been degraded, particular restoration activities can take on especially privileged, sacred meaning. In salmon restoration, for example, the collection, fertilization, and incubation of eggs, and the killing of the female salmon that such necessitates, are seen, paradoxically, as especially meaningful, even sacred, acts. For without the successful propagation and reintroduction of fingerlings to the river, the salmon native to a particular watershed will become extinct, making it a high-stakes, and an emotionally charged, endeavor for salmon restorationists.

Further heightening the importance of this activity for salmon restorationists is the fact that a female salmon's ripe eggs, which are approximately the size and color of pomegranate seeds, are extremely fragile, particularly when they are first fertilized. In the wild a female salmon lays her approximately four thousand eggs in a nest of stones (a redd), which the male then fertilizes with a white, milky liquid (milt) that is released from a vent opening on its underside and covers the eggs like a cloud. Salmon restorationists attempt to replicate this act of creation, which requires technical expertise, calculated precision, and fastidious monitoring. Each fertilized egg, for instance, needs to be propagated individually, maturing for the first month in a small, mesh basket that rests in a trough filled with water that is pumped and filtered from a feeder creek of the main river.[64]

As with the elements and activities involved in religious rituals, however, the requisite acts of egg propagation can take on sacred meaning for some salmon restorationists. On this, House writes:

> Each of us has performed this rite [of killing the female salmon to collect and fertilize her eggs] a number of times before, but it never ceases to be weighted with nearly intolerable significance, the irreducible requirement to do it right.... Everything is ready. Gary [the fish biologist] measures a few teaspoons of the anesthetic into the tank.... Stevie [a resident volunteer], nearly dancing with anticipation, runs to the holding tank once Gary tells him we are ready, lifts the New Year's female out of the water in her tube, and rushes her down to the drugged water in the stock tank.... I have handed the ironwood club to Stevie.... The club comes around [and] connects solidly at the back of her head, just behind her eyes. She shudders for a moment and is still.... [Gary] hands her to David [volunteer], who waits while I scramble for one of the white buckets. Each of us is muttering cautionary instructions to the others, careful, careful, head down, head up, don't drop her now. No one hears. We have moved beyond our nervous ambivalence at the arrogance of our intention and are wholly occupied by the ritual.[65]

Once the female salmon is cut open, her ripe pink-orange eggs are released into the white bucket, white fertilizing milt of the male is released in with them, and the mixture is stirred to combine the coelemic fluid surrounding the eggs with the milt. Clean water is then dribbled into the bucket through surgical tubing, causing the hole in each egg to close and, within an hour, a thin membrane to form. On this House states: "I lower the fingers of one hand into the heart of creation [the egg, milt mixture] and stir it once, twice. For a moment my mind is completely still. Am I holding my breath? I am held in the thrall of a larger sensuality that extends beyond the flesh."[66]

In this particular restorative act, the "intolerable significance" of doing it right, biologically speaking—capturing the female, releasing the eggs, squirting the milt, mixing the mixture, and so on—takes on symbolic, ritual significance, sacred significance. Ecological act has, paradoxically, become sacred act; scientific work, spiritual work. In this case the paradoxical character of the act hinges on the restorationist's simultaneous participation in instigating death and attempting to regenerate life; in doing something she knows intellectually, scientifically to be necessary and finding herself emotionally, sensually drawn into the heart of creation.

The above perspectives raise questions related to the ways in which the notion of the sacred is understood in relation to environmental damage and healing. As I have noted previously, many restorationists tend to assume that nature's authenticity is located within nature itself and is diminished or lost through human influence. According to this view, an ecosystem's original, real authenticity is ultimately irredeemable by human activities; it is lost by direct or indirect human influence, never to be fully recovered by human hands again.[67] This is the type of argument many environmentalists have tended to use in relation to damaged and restored ecosystems: Restored nature is just not as good as the real thing. If given the choice, for example, Holmes Rolston, despite his support in general for restoration activities as we have seen, states that he would rather hike in an old-growth forest than a restored one any day. Restored nature is not as ecologically rich or diverse, neither does it hold the type of historical character as original, say, old-growth nature.

Again, arguments such as these rely on the assumption that nature's authentic, really real, sacred value is given, discovered, or revealed within nature itself. Alternatively, however, some restorationists, such as William Jordan, argue that the sacred value of ecosystems is created by particular human actions. This is what we find, writes Jordan, when we look to "the earth-based religions of indigenous people, which are at least most obviously related to the work of restoration."[68] Drawing on religious historian Mircea Eliade's notion of archaic ontology, Jordan proposes that we understand restoration as an act that can create the real, authentic quality of ecosystems by consecrating them and in effect making them worthy.

"Things—including places or landscapes—are not real in [the premodern or indigenous] view until they come to participate in a transcendent reality," writes Jordan (following Eliade).[69] "Things are not merely found or discovered to be real or sacred, but rather are made so by . . . human acts, which . . . repeat the [primordial] act of creation."[70] When restorationists attempt to copy or replicate a natural system's parts, processes, and functions, they are in effect repeating such an act—they are attempting to participate in a basic act of natural re-creation. According to this view, nature's ontological, essential value is not given, revealed, or discovered; it is "actually dependent on deliberate human acts" of creation. The restorationist ontologically redeems nature's value when she takes it up in her hands and attempts to copy it, to replicate its natural ways of re-creation.

But does nature need humans, or the divine for that matter, for its redemption? My own view is that it does not. Further, my perspective is that such conceptions of spiritual ecological redemption risk promoting an overly anthropocentric view of the human relationship to nature on the one hand (e.g., humans as divinely mandated managers of creation) and anti-evolutionary view on the other (e.g., earth's redemption as a supernatural phenomenon and/or otherworldly reality). The language of redemption may, in some sense, provide a meaningful interpretation of restoration activities, a topic I address in the following chapter. For now, suffice to say that in my estimation nature's true or really real value resides in its inherent capacities for self-renewal and regeneration, or its prolifically wild, generative character, perhaps even beyond human knowledge. Along these lines nature's real value is neither wiped out completely when damaged by human activities nor in need of human (or divine) redemption to reactivate its wild processes.

Ecosystems may require human acts of regeneration or restoration to once again become the thriving natural mediums through which human spiritual and moral meaning may be formed. Though once processes and functions are jump-started by humans, the "power belongs to Nature," as Mills writes.[71] Restoration, state the Chicago Wilderness restorationists, is ultimately "not about control, but surrender" to nature's ways.[72] This understanding follows one of the basic claims of restoration: that it is attempting not to create an entirely new ecosystem based on the fleeting interests and whims of humans but rather to regenerate the evolutionary historical trajectory of the ecosystem prior to human disturbance.

Restorationists generally are not interested, for example, in attempting to create a rain forest where there was once a prairie or, more realistically, an oak savanna ecosystem where there was once long leaf pine. And this is not only because certain ecosystems work better in particular locales and climates but mostly because restoration is an attempt to regenerate the evolved ecosystem and beings that were once there. Conditions of ecological change may force restoration to revise its emphasis on natural history, given that certain organisms may no longer thrive, for instance, where an area's climate has become significantly altered. Nevertheless, restoration's respect for the historical ecosystem is based on the idea that the particular landscape once had and still does have, to varying degrees, certain distinctive, resident, locally evolved natural values—for the benefit of humans but also for the other wild beings that once resided there.

Some restorationists, as we have already seen, refer to this regeneration of natural historical ecosystem value as a way of "serving the sacredness within nature" or "touching the sacred within nature with their hands." These restorationists recognize that not all nature's "real," authentic, wild value is completely wiped out when ecosystems are damaged by human activities. The wildness within nature and divine presence within creation will persist, push on, prevail, according to religious restorationists. "If it can be recalled, it may be restored," as Mills writes.[73] As long as there is biological life on earth, sacred seeds of wildness will remain within ecosystems; where wildness has been diminished, it can be resurrected, "within all life places on Earth, including the human soul."[74] The work of healing damaged land can be viewed as sacred work. Sometimes humans can actually reach in and touch the heart of creation.

"REMOVING THOSE DAMN WEEDS": RESTORATION AS PUBLIC WITNESS

Another way in which the acts of restoration become more than "just the work that needs to get done" is that they can enact, indeed attempt to create the way, things ought to be in the face of the way things currently are. Related to this symbolic dimension of restoration work, religion scholar Jonathan Z. Smith writes that "ritual is, above all, an assertion of difference . . . a means of performing the way things ought to be in conscious tension to the way things are."[75] In this way restoration practice may function as a form of public witness to the collective moral failure of industrial society—and a call to do better—to live respectfully and harmoniously with land.[76]

Unlike other forms of environmental and religious witnessing, however, the public witness of restoration occurs most often inherently in the activity of restoring nature itself. It is not the kind of witness usually intended through civil disobedience. For example, the witness to the societal "sin" of the massive destruction of the tall grass prairie ecosystem throughout the Midwest (less than 1 percent of the original nearly 150 million acres remains) comes for restorationists not through hammering on or "monkey-wrenching" the two-ton tractors and combines of farmers who have historically been the ones to plow and till under the prairie. Rather, it comes through the action of regenerating and returning, giving back, prairie to the region.

The five-acre tall grass prairie plot that was put in at my children's elementary school in rural central Philo, Illinois, for instance, makes a small but public statement, a public witness, to the fact that it was not good, is not good—ecologically, aesthetically, morally, spiritually, and economically in the long term—to create a landscape entirely dominated by a monoculture of nitrogen-intensive cash crops. Initiated by a small group of parent volunteers, with the assistance of several state conservation practitioners, the project was viewed as an educational opportunity for students and the community. The school district owned a plot of land next to the elementary school, which high school students from Future Farmers of America (FFA) used as an experimental planting field. The five acres closest to the school was designated as the Unity East Prairie Project and now thrives as a healthy, though small, intact tall grass prairie ecosystem. Maintained—weeded and fired—by parent and student volunteers, the prairie includes a narrow walking path with small identification signs that mark various prairie plant species. Classrooms use it as an environmental educational resource, and students and the community enjoy meandering its paths.

The deeper meaning and ethical import attributed to restoration as a type of witnessing action is one reason why planting the prairie at the school, at least at the project's outset, caused no small amount of consternation in a community, small and farming based, that still in large measure is skeptical about the value of the prairie ecosystem ("We've been trying to get rid of those damn weeds for a hundred years, and now you want to bring them back?"). Some farmers saw this little plot of prairie as a moral indictment of their way of life and work and, more generally, of the American achievement of agricultural progress.

Aldo Leopold spoke to this broader symbolic meaning of restoration in his 1934 dedicatory remarks at the 1,200-acre arboretum at the University of

Wisconsin, the birthplace, as we have already noted, of the modern study of restoration ecology. At the dedication, Leopold famously said:

> If civilization consists of cooperation with plants, animals, soil, and men, then a university which attempts to define that cooperation must have, for the use of its faculty and students, places which show what the land was, what it is, and what it ought to be. This arboretum may be regarded as a place where, in the course of time, we will build up an exhibit of what was, as well as an exhibit of what ought to be. It is with this dim vision of its future destiny that we have dedicated the greater part of the Arboretum to a reconstruction of original Wisconsin, rather than to a "collection" of imported trees. I am here to say that the invention of a harmonious relationship between men and land is a more exacting task than the invention of machines, and that its accomplishment is impossible without a visual knowledge of the land's history.[77]

In a letter to the American conservationist William Vogt, Leopold wrote that "the idea of a restorative relationship with the land is incompatible within the drives of industrial civilization; one insists that we discover and respect the order of nature, the other urges us to triumph over it."[78] Despite his critical view of industrial culture, Leopold himself found ways professionally and personally to live out a restorative relationship with the land, including the restoration project at the arboretum just cited, as well as the one he initiated with his family at his dilapidated farm in Sand County, Wisconsin.

The Benedictine nuns of Holy Wisdom Monastery emphasize the public dimension of their restoration work in terms of "responding to the needs of the time." A half a century ago, the monastery land surrounding Lake Mendota, a filtering lake for the larger Madison area, was extremely eroded from decades of intensive, agrochemical-based farming practices.[79] The surface area of Lost Lake, which lies at the western boundary of the monastery, was reduced to less than two acres because of sedimentation from surrounding farming practices and residential development. Where tall grass prairie had once dominated the Wisconsin landscape (estimated at a minimum of two million acres), serving as a natural filter for the watershed, it was massively diminished to less than thirty-five hundred acres. The land surrounding the monastery had become a "wasteland," as Sister Mary David puts it, contributing to the pollution of the larger watershed.

Recognizing the collective need to regenerate and protect the watershed, as well as the natural lands of the monastery, the Benedictine sisters slowly began working to restore the health and beauty of their home-place's 130 acres. Eighty-five thousand cubic yards of accumulated silt were removed from Lost Lake, and the shoreline was restored with native plants. Restored to near its original depth, the lake now acts once again as a natural deterrent that detains and filters water that would otherwise wash downstream to neighboring properties and Lake Mendota. Prairie restoration activities began in 1996 and continue each year. To date, 95 acres have been restored to upland prairie with donated seed or seed collected by volunteers and college interns. Each year, 10 to 20 acres have been hand sown with a large variety of native Wisconsin prairie flowers and grasses; these plants have long, deep root systems that prevent soil erosion. Finally, a detention basin was created on the eastern side of the property, and a soil berm below the natural grass waterway, serving as additional forms of water filtration and protection for the north shore of Lake Mendota.

Recently, the Benedictine nuns' restorative work was extended to include the monastery's built structures (also a type of restorative action, which I say more about in the next chapter), with the main house being converted to an ecologically sustainable building. The monastery's multiple restoration projects have received numerous awards from area conservation and community-based groups. Cleared and degraded farmland and choked lakes and streams have become thriving prairie, gardens, and waterways. The Holy Wisdom Monastery has become known in the broader Wisconsin area as "an oasis of quiet beauty where all can come and experience God's presence."[80] The restored landscape has become an outward sign of an inward grace for the Benedictine nuns, a sacramental witness to the regenerating power of the divine presence within and around.

A secular example of the ways in which restoration work can become a form of public symbolic witness comes from a local festival in Lake Forest, Illinois. "The Bagpipes and Bonfire festival" centers on the yearly act of burning exotic and weedy nonnative plants that have been removed from natural preserves around the area. Stephen Christy, executive director of the Lake Forest Open Lands Associate, describes the festival, writing, "The festival offers family entertainment, period actors, hot-air balloons, and food and drink. Then, at dusk, a 100-piece Scottish piping band emerges from the prairie, solemnly circles the brush pile, and plays traditional airs. A solo finale of 'Amazing Grace' rolls across the prairie as a torch bearer emerges from the woods far to the west. He runs up the hill, circles the pile three times, and then lights it. As darkness settles, the crowd intently watches the blaze, backing away as its heat increases. The effect is magical, and silence reigns. The cycle is complete, and the fire works its hypnotic, purifying effect on the landscape and people."[81] The festival that began as a small gathering a decade and a half ago has grown into a community tradition drawing more than a thousand people.

Herein lies one of restoration's most significant symbolic aspects: it does in fact create restorative relationships between and among individuals and land, *despite* and *in the midst of* the current "drive of industrial civilization." There still is room, restoration work reminds us, to attempt to respect the order of nature, to learn how to live more harmoniously, more beautifully, more meaningfully with land. People can come to know a particular landed place and be drawn into its slow, self-healing ways. Land, if given the chance, will come back to prolific, thriving, wild life. The human spirit and heart can be transformed and renewed in the midst of fragmentation and degradation. Wholeness can be found within a fragmented land, as nature writer Janisse Ray claims in her narrative account of the restoration of the Pinhook Swamp linking the southern wildernesses of Okefenokee Swamp and Osceola National Forest.[82]

In sum, restorationists most often utilize scientific ecological language to describe the work of restoring damaged land. But they also at times rely on religious and ethical understandings to explain personal experiences of transformation and renewal, as well as the broader symbolic meanings of restoration actions. The practice of ecological restoration serves for some restorationists as sacred work and public witness. In these ways it provides a ritual action for creating deeper values in relation to particular landscapes and communities of people.

More will need to be said regarding the ways in which restoration-inspired spiritual and moral values might relate to those of society more broadly. Furthermore, the question of whether religious understandings can helpfully contribute to public restoration discourse remains to be seen. We turn now to examine these issues, exploring the ways in which restorative symbolic action may help to further form a broader cultural narrative of connection between people and land. For even as restoration may shape personal experiences, collective values, and cultural and religious meanings in relation to land, it may also, in a larger sense, contribute to the formation of a new societal era.

NOTES

Chapter 5 is based on "Ecological Restoration as Public Spiritual Practice," published in *Worldviews* 12 (2008): 237–52. The material is reprinted here with permission.

1. This has generated a large debate within the literature. On the professionalization of restoration, see, for example, Light, "Restoration," in *Restoring Nature*, 163–84.
2. See Higgs, *Nature by Design*.
3. See, for example, Light and Higgs, "The Politics of Ecological Restoration," 227–47; and Van Wieren, "Restoration as Public Spiritual Practice," 237–54.
4. On the idea of ritual in relation to everyday activities, see Gould, *At Home in Nature*, 63–101.
5. Mills, *In Service of the Wild*, 125.
6. Ibid.
7. See Jordan, *The Sunflower Forest*, 37. See Sideris, *Environmental Ethics*.
8. See Jordan, Barrett, Henneghan, et al., "Foundations of Conduct."
9. Jordan, *Sunflower Forest*, 72.
10. Ibid.

11. Jones, *Feminist Theory and Theology*, 150.
12. Ibid., 151.
13. See Jordan, *Sunflower Forest*, and Holland, "Restoration Rituals," 121–25.
14. Jordan, *Sunflower Forest*, 5.
15. Bron Taylor, *The Encyclopedia of Religion and Nature*, s.v. "Restoration Ecology and Ritual."
16. Jordan, *Sunflower Forest*, 53.
17. Quoted in Mills, *In Service of the Wild*, 127.
18. Ibid., 31.
19. On this idea, see Rolston, "Naturalizing and Systematizing Evil," 67–86.
20. Ibid., 32.
21. See Rasmussen, "Returning to Our Senses," 40–56.
22. See Santmire, *Ritualizing Nature*.
23. "Modern Yoga," states de Michelis, is "a technical term to refer to certain types of yoga that evolved mainly through the interaction of Western individuals interested in Indian religions and a number of more or less Westernized Indians over the last 150 years." Thoreau, according to de Michelis, appears to be the first Westerner to affirm himself (in 1849) as a yoga practitioner. See de Michelis, *A History of Modern Yoga*, 2, 60, 251.
24. See my "Ecological Restoration as Public Spiritual Practice."
25. Atler, *Yoga in Modern India*, 239.
26. Strauss, *Positioning Yoga*.
27. De Michelis, *History of Modern Yoga*, 251.
28. Benedictine Women of Madison, "A Living Tradition of Care," www.benedictine women.org/care/care_envhistory.html (accessed May 28, 2010).
29. Meisel and del Mastro, *Benedict*.
30. Personal communication, May 28, 2010.
31. Ibid.
32. This quote is by a "volunteer in community," a program at the monastery where single women come to live and work in Benedictine community for two to four weeks during the summer. These volunteers typically work four hours a day on the monastery's natural lands. On this, see Benedictine Women of Madison, "Volunteers in Community," http://benedictinewomen.org/2012/volunteers-in-community/ (accessed May 3, 2012).
33. Personal communication, May 28, 2010.
34. Benedictine Women of Madison, "Environmental Education," http://benedictine women.org/care-for-the-earth/natural-environment/environmental-education/ (accessed May 1, 2012).
35. Personal communication, May 28, 2010.
36. Benedictine Women of Madison, "Nature Trails," http://benedictinewomen.org/care-for-the-earth/natural-environment/nature-trails/ (accessed May 1, 2012).
37. Benedictine Women of Madison, "Nature Notes," http://benedictinewomen.org/care-for-the-earth/natural-environment/nature-notes/ (accessed May 17, 2012).
38. Ibid.

39. Benedictine Women of Madison, "Connect with the Dynamic Beauty of Nature," www.benedictinewomen.org/care/care_volunteer.html (accessed May 28, 2010).
40. Personal communication, May 28, 2010.
41. Ibid.
42. Stockinger, "Classical Music Review."
43. Ibid.
44. Catherine Bell uses the term "ritualization" over "ritual" given the former concept's attention to the active process that occurs when people privilege one type of activity over and against another. See her *Ritual Theory, Ritual Practice*, 32–35.
45. Gould makes a similar point in relation to homesteading. See her *At Home in Nature*, 63–101.
46. Mills, *In Service of the Wild*, 144.
47. See Miles, Sullivan, and Kuo, "Psychological Benefits of Volunteering," 218–27.
48. Personal communication, May 28, 2010.
49. Benedictine Women of Madison, "The Divine Is Present in All Creation," www.benedictinewomen.org/care/care.html (accessed May 28, 2010).
50. Personal communication, May 28, 2010.
51. Benedictine Women of Madison, "The Divine Is Present in All Creation."
52. Personal communication, May 28, 2010.
53. Mills, *In Service of the Wild*, 2.
54. Ibid.
55. Early on, Thomas Berry claimed a bioregional social philosophy as integral to an earth-based spirituality and ethic. See his *The Dream of the Earth*, 171–79.
56. Personal communication, May 28, 2010.
57. Ibid.
58. House, *Totem Salmon*.
59. Ibid., 6.
60. Ibid., 8–9.
61. Ibid., 18–19.
62. Ibid., 19.
63. Personal communication, May 28, 2010.
64. House, *Totem Salmon*, 100.
65. Ibid., 101–3.
66. Ibid., 104.
67. On this point, see Jordan, "Ghosts in the Forest," 3–4.
68. Ibid.
69. Ibid., 3.
70. Ibid.
71. Mills, *In Service of the Wild*, 2
72. Ibid.
73. Ibid., 10.
74. Ibid., 9–10.
75. Smith, *To Take Place*, quoted in Gould, *At Home in Nature*, 63.

76. On the notion of public witness within environmentalism see Gottlieb, *A Greener Faith*, 166–67.
77. Flader and Callicott, *River of the Mother of God*, 210–11.
78. See Worster, *The Wealth of Nature*, 180. This quote is environmental historian Donald Worster's interpretation of Leopold's letter.
79. On the information cited in the following paragraphs, see the website of the Benedictine Women of Madison, "A Living Tradition of Care."
80. Ibid.
81. Holland, "Restoration Rituals," 123.
82. Ray, *Pinhook*.

CHAPTER SIX

◌⁓

RE-STORYING EARTH,
RE-STORIED TO EARTH

It's all a question of story. We are in trouble just now because
we do not have a good story. We are in between stories. The
old story, the account of how the world came to be and how
we fit into it, is no longer effective. Yet we have not learned
the new story.

—Thomas Berry, *The Dream of the Earth*

In his classic book, *The Dream of the Earth,* Thomas Berry proposed that we are entering a new phase in human history: the ecological age. This age, different from previous historical eras such as sixteenth-century scientific discovery or eighteenth-century industrial expansion, would be marked by a "vision of a planet integral with itself throughout its spatial extent and its evolutionary sequence."[1] Beyond simply reducing fuel use, modifying economic controls, or slightly reforming our education, the realization of this vision would require human psychological and cultural transformation on a planetary scale. It is not simply a matter of reducing our fuel use, modifying economic controls, or slightly reforming our education system, wrote Berry. "Our challenge is to create a new language, even a new sense of what it is to be human."[2] What we need is a new story, Berry proposed, a new narrative and deepened understanding of our relationship with the community of life on earth.

Many secular and religious environmentalists have, over the past twenty-five years, heeded Berry's call to develop new stories about the relationship between humans and the earth. Authors have proposed plural versions of this new, ecological story, including, for instance, re-storied narratives about what it means fundamentally to be human, as well as narratives about the place and role of the divine in relation to earth.[3] This book too has suggested implicitly that a new story, or new stories, about nature, humanity, and the sacred are needed for contemporary society, particularly as the stories relate to concrete actions to heal earth's natural

170

systems and communities. In this final chapter I want to make this claim more concretely, arguing further that human society is, in fact, presently entering a new era—namely, an era of restoration.

Beyond entering an ecological age in general, I propose that society is entering one particularly marked by the deliberate creative and systematic attempts of humans to restore connections with the natural world. As I stated in this book's preface, the time has come for environmental ethics and society to move toward more positive, forward-leaning approaches and solutions to environmental issues. This is not only because a crisis-oriented approach has not finally worked in creating more cooperative relationships between people and land but also because emphasizing such an approach overlooks the myriad of examples, and stories, of ecological healing and restorative action that are emerging.

Ecological restoration and its practices such as I have discussed throughout previous chapters is not the only form of healing activities that are already giving shape to a restoration age: for example, the restorative environmental building, urban farming, and children and nature movements too are contributing. In support of this proposal it is necessary to explore the new story, or multiple, though overlapping stories, that are integral to the broader project of restoration that is unfolding in contemporary society. For it is not that ecological restoration, or other restorative types of activities, need a new story per se, rather that further developing, integrating, and telling these stories that already exist is important to further develop modern humanity's relationship to the natural world.

Emphasizing the notion of re-storying the relationship between people and nature, I begin by examining the distinctive functions of story and narrative in bringing about a new ecological consciousness and culture. Next I look specifically at the role of religious ethical narratives in the re-storying of the relationship between people and land. I then propose an agenda with six themes that are arguably critical for the development of a broader cultural story of restoration. In conclusion, I consider the overall thesis of this project and a final note on the emergent and dynamic character of land-based narratives.

STORY AND ENVIRONMENTAL CHANGE

Although other environmental writers have proposed that a new story about the relation of humans to earth is needed in order for significant cultural and therefore also environmental change to occur, it is important to examine some of the *explicit* connections between story and environmental consciousness and culture. For understanding these connections more concretely may lend insight to the ways in which story relates to the formation of particular patterns of action in relation to the human-nature connection. There are at least three ways in which we might understand the relation between story or narrative and environmental change.

First, symbolic-cultural narratives contribute in fundamental ways to the formation of human identity. More particularly, narratives of connection between people and land are integral to the formation of ecological identity, that is, a sense of selfhood continuous with the natural world. Authors from numerous disciplines have emphasized the ways in which a sense of ecological identity is significant for fostering positive environmental change. In large part who we are and what we strive to become are based on stories from childhood and through adulthood that constitute a framework within which we perceive and interpret our experiences.

For example, recent studies in the field of child and environmental psychology find that children's experiences of the natural world are significantly related to the development of moral ecological consciousness and care for particular animals and places.[4] Additionally, religious stories frequently speak to the basic elements of human existence, whether in family or in society, offering templates or horizons that importantly determine how people make sense of, give meaning to, the worlds around and within them. This is one reason why telling and enacting sacred stories to and with small children is often a central part of religious education; because the stories of particular communities, whether religious or secular, have historically played an integral part in helping persons interpret their roles and place in relation to others and to a collective whole.[5] As religious ethicist Anna Peterson writes, "only in light of stories can people come to understand themselves, the multiple roles they play, and the origins and trajectories of their communities."[6]

This does not mean story inevitably or unavoidably connects us to our histories, or our ecologies, or that it does so with any accuracy or adequacy. No matter what, however, stories of our particular life-places provide room to navigate, adjust, reject, affirm, or change the life and world that are ours. As Carolyn Merchant writes, "seeing ourselves as a part of a story in which we play a role guides our actions; the storyline often tells the actor what to do or conversely allows an individual to rebel and follow a different story."[7] Along these lines adopting or creating a new story for oneself and for broader society can play a role in promoting individual and social transformation.

Second, stories can engage and spark the imagination in ways that connect people to particular natural and cultural locales, and foster avenues for personal and social change. In his insightful book on bioregional thought, *LifePlace*, landscape architect Robert Thayer writes about the role of the imagination in living in a place well. "Without imagination, humanity mires in mediocrity and stagnation; to imagine, to create is to survive and thrive."[8] Various artistic media—visual art, drama, dance—can promote the imagination in relation to particular places, but storytelling does so in unique ways. Humans, as intelligent, social creatures, are also narrative-building creatures. We come to know and understand ourselves through language and symbols, thereby orally describing and giving meaning to our life and our experienced realities. Telling stories about natural landscapes and familial and social communities allows expressions of what these phenomena and relationships really are.[9] This, in turn, connects us to them, for better or worse, in deep and lasting ways.

This leads to a third point regarding the relation of story and environmental change. That is, new cultural narratives are often central in fostering transformative social movements. The civil rights movement and the contemporary environmental movement, for example, have historically gained momentum in times when new narratives about human beings or social or natural systems have increased in prominence precisely because of their capacity to describe more adequately the experiences of people in relation to their everyday realities. In the 1960s Rachel Carson, for instance, told a "new" story about the role of humans and the natural world in her alarming book *Silent Spring*. Carson's account of the disastrous effects of the pesticide DDT on populations of American songbirds generated a larger ethical narrative about the potentially destructive ecological and social impacts of modern technology and its use in agricultural activities. The environmental movement was spawned, in part, in response to this new narrative about the potentially harmful role of human practices and the development of modern technology in relation to the natural world.

A final note about story and environmental change is needed. That is, narrative is a genre subject to interpretation, alteration, and adoption in ways that are influenced by particular contexts and experiences. Even a new grand narrative of humanity and nature, needing to take into account the evolutionary historical story of the creation of the universe, will not necessarily be interpreted or articulated in the same way by all humans or even by all people in any particular community. Merchant, for example, calls for an overarching narrative of partnership between humans and nature. Nonetheless, she accepts the inevitability of multiple perspectives, interpretations, and expressions of any new partnership story. The implication is that these will serve a common purpose but from a plurality of perspectives. "Environmentalists, feminists, and philosophers all call for different pathways to sustaining the earth," states Merchant.[10] "Many narratives stemming from many groups of people worldwide may contend with each other or at some point eventually come together. The new voices and stories that contribute to a sustainable world will have new visions of belonging to the earth."[11]

The re-storying of the relationship between people and land will indeed involve multiple, and at times conflicting or overlapping narratives of environmental, social, and religious restoration. The ways in which these restoration narratives may be woven together over time represents a work in progress. To this end I propose an agenda below that may help guide the re-storying process as it unfolds. One more preliminary step is needed, however. The role of religion in relation to the re-storying of people and nature needs probing. For just as it is important to understand the connection between story and environmental change more explicitly, so too is it extremely important to understand the relationship between religious narratives and a new restoration story.

REDEMPTION, RELIGIOUS ETHICAL NARRATIVE, AND RE-STORYING

Environmental authors from various quarters have indicated problems with considering religious appeals in relation to environmental ethics and civic discourse about environmental issues.[12] In his book *A Greener Faith: Religious Environmentalism and Our Planet's Future* philosopher Roger Gottlieb highlights five such problems. One problem, for example, is that religious beliefs are irrational, or at best nonrational, and thus have no place in the organization of society. Taking it further is the sentiment that religious values are peripheral to environmentalism, which should be shaped by science, not faith.[13] While Gottlieb sees significant problems with religious appeals in public environmental discourse, he nevertheless does not find them insurmountable. "I am confident," he writes, "that secular culture, including politics and science, can learn a great deal from religious traditions and temperament. For all their limitations, religions still have a great deal to offer."[14]

As I have stated at various points in this project, I too share the concerns regarding religious appeals that authors such as Gottlieb have raised, but I also do not think they are insurmountable. In fact, I have suggested that appeals to a religious dimension of the human relationship to nature may serve the restoration movement well. Further, I have proposed that religious and scientific perspectives can converge, or at least complement each other, in the context of environmental practices, such as restoration, that involve scientific and experientially based types of understandings. Here I would like to explore more explicitly some of the ways religious ethical narratives might contribute to a broader story of restoration. To do this, I draw on a single, though prominent, religious narrative, one that is also integrally related to the idea of restoration—that of redemption.

The Problem with Redemption

Although it may appear obvious to draw on religious language of redemption to interpret ecological restoration, I have reserved this discussion for the final chapter for two reasons. First, recall this book's basic premise and argument that environmental ethical values, norms, and ideals require grounding in vernacular social ecological actions for them to take root and flourish in people, communities, and society. Setting aside an examination of the notion of redemption until now allows it to be more fully shaped by the actual restoration-based experiences and values that we have explored in previous pages.

Second, some ecological theologians have critiqued traditional Christian understandings of redemption for their individualistic and otherworldly bias, making it perhaps not as obvious, or at least not as appealing a choice for interpreting ecological restoration's religious meaning. Too often, as Daniel Spencer writes, Christian notions of redemption and restoration have "come at the expense of the earth: either the earth's integrity is absent or ignored and thus serves merely as a passive stage on which the post-Eden drama of human salvation-restoration [redemption] is played out, or it is sacrificed in apocalyptic, end-of-the-earth scenarios where the destruction of the planet is a necessary prelude to Christ's return and saving action."[15]

Additionally, classical Christian narratives of redemption have historically emphasized individual human redemption to the neglect of nonhuman others and the creation as a whole.[16] Moreover, as Thomas Berry and others have argued, the Christian notion of redemption has been wedded with understandings of divine providence and material progress, especially in the American context, making it particularly susceptible to theologies of exceptionalism and manifest destiny.[17] Related to this critique and Spencer's above, some ecological theologians have worried that continued use of the theological concept of redemption in relation to land will be interpreted in terms of the restoration of a New Eden on earth or, otherwise, the restoration of pristine wilderness-like ecosystems where humans are considered (relatively unwelcome) visitors rather than active ecological residents.[18]

This said, redemption, for better or worse, remains a central motif within Christianity and Western culture.[19] For this reason some recent ecological theologians have attempted to reinterpret the Christian metaphor of redemption in relation to the whole of creation, rather than only to human beings, as a way to redeem its value in light of evolutionary ecological understandings.[20] Relying on biblical passages and theological motifs that support a cosmic view of Christ's activity in the universe, arguments are made to the effect that God created the whole earth essentially good, so it too is fit for and part of God's saving action in Jesus Christ. According to this ecologized interpretation of redemption, all creation, here and now, groans for God's saving activity (Romans 8). Further, in the final, heavenly eschaton yet to come, no part of creation will be accursed any longer; the lion will lie down with the lamb, death for all creatures will be no more, all shall be made new (Isaiah 11 and 65; Revelation 5 and 22).

Even as these interpretations of redemption provide an important step for their ecologizing, they nevertheless also point to potential difficulties, especially where restoration practice is concerned. First, as already noted, restoration raises the question of whether nature, damaged by past human actions, is in need of redemption in a theological sense. Redemption in much of Christian thought involves the belief that God in Jesus Christ in some way compensates or makes amends for (and thus saves) an inherently finite, fallen, and sinful humanity.

Additionally, in the prophetic tradition, redemption is understood in terms of God's vision of shalom in which the people of Israel are divinely delivered and liberated from the social injustices of exile and slavery.[21]

Both of these theological interpretations, however, make one wonder, as I did in the previous chapter, whether it is really evolutionarily, ecologically adequate to say that ecosystems, where damaged by humans, are in need of divine redemption or liberation? Further, in Christian conceptions of redemption, humans are in need of redemption because of their unavoidable fallen condition, an intrinsic lack in existence (sin), however this is characterized. But humanly degraded nature (or healthy nature) is neither inherently fallen nor existentially lacking in the same way as humans may (or may not) be. Moral restitution might be in order for land communities and species that have been wrongly damaged by unjust, callous, or ill-conceived human acts, as I have argued throughout this project, but not divine redemption.

To resolve these difficulties some ecological theologians have found persuasive an idea of co-redemption whereby God and humans work hand in hand to bring liberation and healing to an otherwise impoverished and degraded earth.[22] Larry Rasmussen, for example, has utilized a variation on this understanding of co-redemption, stating: "Redemption means reclaiming broken or despoiled or unfinished creation for life. It is not extracreational or extrahuman, much less extraterrestrial. Redemption consists in Spirited action, often very ordinary everyday ones, against the anticreational forces that violate creation's integrity and degrade and destroy it. The ultimate goal is 'a new heaven and a new earth.'"

Yet restoration is a human action, with human goals, to make amends for previously destructive human actions. Christian communities may want to say that living out a life that has been redeemed (or restored) by God consists of performing everyday actions against nature's degradation and for its restoration. Nevertheless, stating that religious restorationists are really doing God's redemptive (or restorative) work on earth, or that restoring this or that ecosystem is part of God's divine will, seems to smack of the hubris and presumption that philosophers such as Katz and Elliot originally worried about in relation to restoration activities.

Rather than asserting their will or being certain they are doing the right thing in relation to natural ecosystems, restorationists, as we have already seen, often attempt to quiet their own will or ego in order to objectively and attentively observe nature's own processes and functions. It may be fitting to say that Christian communities are performing a type of faithful service, or contemplative, prayerful action, in working to reverse a tide of human-induced degradation and allow nature's voice to once again express itself. Or it may be that Christians need to experience ecological restoration in order to rightly understand theological meanings of human redemption. But nature itself is not in need of divine redemption, and humans should not presume they are doing God's redemptive work when they perform acts of restoration; best, in other words, to leave redemptive language out of it.

Re-Storying Redemption

In response to the difficulties with redemption just cited, there is an alternative way that religious language about redemption and forgiveness can be narrated. That is, redemptive language of healing and forgiveness can be used to describe the spiritual ecological experience. Christian communities working restoratively, for instance, may narrate the ways in which experiences of divine presence are opened in and through the collective process of working to bring naturally evolving life back to land. Restorationists may describe experiences of forgiveness, grace, and healing by "living with" and observing land's self-healing capacity and emerging, dynamic beauty. Through regenerating degraded natural landscapes, one might experience regeneration, a biological and theological term, within one's own self and within the community.

In this way an ecologized notion of redemption is narrated from within a broader context of land health. Furthermore, on this understanding, the experience of ecological restoration helps religionists to more expansively understand theological meanings of redemption. Instead of widening understandings of redemption to include nonhuman creation, however, the idea and interpretation of the experience of human redemption is expanded and situated within a notion of creation's integrity. Where ecological degradation reigns, unchecked by human acts of restoration, modes of divine presence, forgiveness, grace, and healing are stifled. Where land is actively restored, communities of people are drawn into the slow, healing power of God's redeeming embrace. On this interpretation of redemption, forgiveness from God is known and experienced in and through experiencing forgiveness from land. Christians "are offered forgiveness," in other words, "as the birds return with an olive branch and the waters stir with life once more."[23]

Additionally, the idea that persons may receive forgiveness from land may refer to the ways in which the restorationist, and society more broadly, can experience a sense of atonement or absolution of guilt for ecological sins through restorative actions. Insofar as land begins to regain its health and diversity, one experiences a sense of forgiveness, a pardon or lessening of guilt for a violation of nature's intrinsic integrity. The restorationist herself may not have directly caused the environmental degradation. Nonetheless, where she recognizes and restoratively lives with the land community in which she finds herself, she, at some level, comes to feel its loss.

The restorationist also experiences the way in which land, where given the chance, can over time heal itself, recovering in a way from the memory of being harmed. On the one hand, particular landed places will never be the same after a rupture has occurred in their evolutionary-ecological historical trajectory. On the other hand land, in regaining its capacity to renew and heal itself, is moving on; it is acting as if it were, in a sense, freed. In turn the restorationist, and society, figuratively speaking, may breathe a sigh of relief. Land has, in a sense, forgiven us for doing it harm. It has moved on to live its own good kind of life—and so can we.

The Sabbath tradition in the Hebrew Bible presents a view of the ways in which forgiveness from God is experienced through the healing of land. The practice of sabbath, or the observance of a day of rest on the seventh day in a seven-day week, was recorded in the Torah as part of the laws given to Moses. Based on the theological idea in Genesis that God rested on the seventh day of creation, the requirements of sabbath law state that no work should be done by any person or livestock on the seventh day. The Sabbatical Year specified in the book of Leviticus expands the weekly idea of Sabbath to a yearly principle, where the seventh year is understood as a year of resting and release—for people, including slaves and hired laborers, and for the land. The Jubilee Year, also written about in Leviticus, further extends the Sabbatical Year, specifying that every seventh Sabbatical Year, the land should be returned to its original owners and that indentured servants should be freed.

Time and again the Hebrew scriptures narrate the ways in which the potential for wholeness, for a right relationship between God and Israel, lies in the relationship between Israel and the land. The Sabbatical laws in particular serve as a guarantor of right relationship between the land community and Israel, and thus Israel and God.[24] In the Holiness Codes, for instance, God instructs Moses on Mount Sinai regarding the ways in which the land ought to be treated once Israel enters the land: "In the seventh year there shall be a Sabbath of complete rest for the land, a Sabbath for the Lord: you shall not sow your field or prune your vineyard. . . . You may eat what the land yields during its sabbath—you, your male and female slaves, your hired and your bound laborers who live with you; for your livestock also, and for the wild animals in your land all its yield shall be for food" (Leviticus 25: 4–7).

Here we have the idea that acts of liberating the land and its inhabitants are integrally and necessarily connected to Israel's experience of a forgiving, faithful God.[25] In the context of restoration, Christians might say analogously that bringing land back to health, giving it a rest, a break, in a sense, from being used and even degraded, is integrally connected to the ways in which they may experience life with a forgiving, faithful God. Being drawn into land's slow, self-healing ways might come to intimate the process of being drawn into divine embrace.

The experience of forgiveness in the Christian tradition is further narrated in the collective ritual of public worship. While forgiveness is understood in personal terms on one level, it is also viewed in communal terms. This is seen most notably in the public confession of the Sabbath liturgy. In many Protestant churches, for example, confession and absolution during collective worship represent the sentiment that all people are sinners, all are finite, all have failed, and likewise all are forgiven, all are redeemed, all are made whole by the grace of God in Jesus Christ.

Restoration too offers an individual and collective experience and expression of forgiveness from land. Recall the reworking of Leopold's dictum I proposed earlier: "To live with an ecological restorationist's education is to choose to live with a group of people in a world of wounds in an attempt to heal them within and without." Here it might be further revised to read: "To live with an ecological restorationist's education is to choose to live with a group of people in a world of wounds in an attempt to heal them within and without—and, in the process, experience just how forgiving, how gracious land, humanity and God can really be."

Religious Ethical Narratives and Re-Storying

As I stated at this chapter's outset, a new restoration story of redemption may also be able to contribute positively to public restoration discourse. Here using the ecologized version of redemption just proposed as one example, I suggest three further ways in which religious ethical narratives might relate positively to the broader re-storying of people and land. These proposals will be much too brief and general, but they can begin to open a dialogue on the intersection of religious and environmental narratives in public restoration discourse.

First, religious ethical narratives such as redemption may offer responses to the question: Why should we engage in healing environmental actions to begin with? In the case of ecological restoration the question becomes: Why restore damaged ecosystems at all? Scientific ecological understandings offer technical advice regarding restoration practices, but religious meanings of restoration may tell us why these should be undertaken.

Perhaps humans ought to restore an environment because the earth and its creatures are intrinsically valuable and sacred, and restoration is a form of service to Other nature (see chapters 2 and 3). Or perhaps people practice restoration because in American history the scythe really did go too far, and restoration is a type of devotional practice that can help heal the wounds of past ecological sins (see chapter 5). Or perhaps ecological restoration can serve as a form of justice-making and community revitalization, a public witness to the ways in which good ecological community can be enacted in today's hyperindividualistic and globalized society (see chapter 4). Finally, in accordance with the above redemption narrative, perhaps humans should restore earth's processes and systems because their spirituality, the fulfillment of relational humanness, depends on the existence of a thriving and regenerating natural world (see chapter 3). Each of these religious ethical reasons may prove beneficial in building new stories of the ways in which people may live restoratively with land.

Second, religious ethical narratives about the human connection to nature may yield further narratives that are encompassing, holistic, and, at times, countercultural in terms of articulating a vision for how people can live in good or right relationship with others, the divine, and the whole of creation. On the one hand, this quality can make religious environmental ethics appear, at times, abstractly utopian or "pie in the sky," detached from concrete, practical environmental problems. On the other hand, the holistic narrative aspect of religious ethics contributes to its distinctive capacity for responding to the large questions of the human relationship to a wounded and healing earth community.

Religious ethical narratives of redemption, for instance, may help illuminate the ways in which human and ecological injustices, and acts of restoration, are often intimately interconnected (see chapter 5). Additionally, religious narratives can accentuate the inherent difficulties and sacrifices and potential fulfillment and joy that come through living in real community with others, as well as the place and role of the sacred in relation to healing. Further, insofar as religious narratives are born of healing actions and experiences, as proposed in this project, they become shaped by the concrete realities of daily life in ways that help guard against overly romantic conceptions of community life. In relation to restoration this may involve narrating visions of what restorative communities might look like in practice, such as the one presented in the previous chapter.

This leads to the third dimension of religious ethical narratives in relation to the re-storying of society noted here. That is, religious ethical narratives may speak in a distinctive voice to the ways in which the relationship between humans and the land community necessitates mediating actions. Throughout this project I have noted the ways in which restoration provides a context for reflecting on the ambiguities in our relationship to the natural world. Restoration practice, for instance, puts individuals face to face with land's myriad processes, functions, cycles, and events in ways that can draw out a wide range of human experiences and emotions in relation to land. These aspects of relationship are dimensions of human life that religious ethicists have thought long and hard about, as I have indicated in various chapters. And although traditional theological ethics has emphasized the ambiguities involved in human-human, human-nonhuman, and human-divine relations, religious communities engaged in restoration work give religious ethical language to these ambiguous aspects in new ways.

As noted in the previous chapter Benedictine restorationists describe their work as a type of sacramental practice that enables people to mediate meaningfully their relationship with other people and creatures, the divine and the natural world as a whole. They narrate the way in which restoration practice brings people together in hands-on activities around the shared effort to heal land, even where an activity may be difficult physically, socially, and emotionally. They speak about the ways in which informal liturgical actions, such as a prayer for the healing of the earth or a shared meal during restoration work, can at times provide symbolic meaning that helps to mediate and deepen the relationship between groups of people working together to regenerate land and community.

These are only a few of the ways in which religious ethical narratives may contribute beneficially to the cultural re-storying of our relationship with particular land communities. Other ways will need to be articulated as well. As noted above I have assumed here, as well as throughout this project, that religious narratives about humans and nature will need to be shaped by ecologically healing action and experience in order for them to relate meaningfully to public restoration discourse. Not all religious ethicists will want to place as high a value on the role of environmental experience in formulating narratives about nature, humanity, and the sacred. The focus of this particular project, nonetheless, has been on restoration experience and the positive values that this experience may generate within communities and society. Additional nature-oriented experiences and sources will need to be considered in future projects related to the restoration age that is unfolding. For now, however, we need to identify an agenda for the restoration age, one that can guide the ongoing effort to re-story the relationship between people and earth.[26]

AGENDA FOR A RESTORATION AGE

Numerous stories of restoration of relationships between people and nature are presently [means "in a little while" rather than "at the present"] unfolding in various spheres, even in today's highly technological and urbanized society. I have attempted implicitly to tell one of these stories, or variations on a theme of a narrative—from wounded land and spirit to healing land and spirit—in this book, as I stated in the preface. I have used ecological restoration practice as a context for narrating a story of restoration precisely because it provides a paradigmatic example of the type of activities and stories required for the development of a broader restoration age. More stories, or necessary characteristics of the content of future stories, are needed for a more full-fledged narrative of restoration between people and nature to become functional within society. Here I identify six characteristics as shapes of an agenda for a restoration age. These characteristics are neither definitive nor exhaustive, but they do set a trajectory, highlighting key areas of concern calling for attention if a new restoration age is to unfold.

First, stories will need to narrate examples of ecological degradation and recovery on a case-by-case basis and then weave them together into a broader narrative, so that wounding and healing actions are linked to particular communities yet exhibit a common story. Stories of people healing earth and our relationship to earth ultimately offer forward-looking, hopeful plots with characters enacting transformative environmental actions. Nevertheless, any story of restoration must begin with the acknowledgment that the natural world and our relationship to it has become degraded in numerous and varying ways, depending on our particular contexts. Environmental narratives most often use language about the environmental crisis or earth's distress or human alienation from nature writ large. Although this may have had its place historically in motivating public concern about the environment, the restoration age requires concrete, vernacular attention to the natural processes and living connections that have been impoverished in particular locales. Healing actions cannot be motivated or developed only by universal proclamations that earth and humanity are wounded. The degradation of biotic communities and impoverishment of human communities is different depending on one's natural and cultural landscape. In the cases of restorative communities described in previous chapters, for example, stories of degradation and recovery, loss and healing, varied when narrated from the coast of New Orleans, meadows of Wisconsin, and the forests of Vermont; yet the plots of these stories of restoration overlapped. Powerful new stories must tell specific, variegated narratives that resonate with a common restorative theme.

Second, the restoration age needs narratives that bridge grieving for what has been degraded with celebration for what has been and can be regenerated—ecologically, socially, and spiritually—in regard to the natural world. Throughout this project I have noted the distinctive characteristic of restoration to orient participants and the wider culture to the creative, paradoxical place between what is a new form of the *via negativa* and *via positiva*, that is, between lament for what has been unjustifiably taken or destroyed and gladness that all is not lost, for it is still possible to give back to land its own power for transformation and renewal. Restoration's truth–claims stem, in part, from the dual recognition that land and spirit have been and continue to be wounded and that restoration activities really can contribute to the regeneration of healthy land, communities, and individuals. The re-storying of our relationship with nature requires stories that tell the whole plot from the perspective of multiple characters inhabiting various locales. These stories must include narratives recalling what has literally, concretely been harmed—who or what are the villains—what can be legitimately, practically regained—and who or what are the heroes.

Third, the restoration age will need narratives that stress the ongoing, dynamic, evolving encounter between humans and the natural world. Restoration of earth is not a final, fixed, static state achieved once and for all when ecosystems, communities, and people have been rejuvenated. Restoration practices may best be narrated as a restorative mode for human life in relation to nature in which people and natural processes and systems become interwoven through ongoing restorative acts.[27] Culture and nature coevolve, or devolve, as they act in more or less mutually restorative ways; we therefore need narratives about the ways in which restorative activities weave together human and biotic communities, currently and historically, in ways that are ongoing and variegated. One way to narrate this concept of coevolution is to tell stories of connection between people and land that incorporate the full human life span. The restoration age will require narrative accounts about the connection between humans and the natural world, beginning with stories of connection formed in childhood and continued into adulthood and old age. Stories for the restoration age will need to involve plots that narrate the ecological life histories of characters, ranging from infancy through death.

Fourth, the restoration age will require narratives about the multidimensional character of the relationship between humans and earth, including its more ambiguous dimensions. Humans are complex emotional, intellectual, spiritual, and physical beings with a range of needs and desires that become activated and fostered in unique ways through our connections with nature. Stories of restoration require rich and complex narratives that spotlight people who speak of the panoply of human capacities that may be developed insofar as the relation of humans and earth is restored. Humans may have an inherent need to live in meaningful relationship with nature, as E. O. Wilson and Stephen Kellert have argued in relation to the theory of biophilia, but part of this relationship, as I have indicated at various points already, is fraught with difficulty, practically and emotionally. This makes for high drama, the stuff of powerful stories.

We may recall characters such as Job in the Hebrew Bible and Ishmael in *Moby Dick* as figures that have narrated multifaceted plots holding the depth and range of emotions that come into play in the encounter between humans and nature. Here we have characters that powerfully articulate a full range of emotions from the tragic to the romantic in response to the natural world, the divine, and the cosmos. Stories such as these retain their power through the ages precisely because they speak to fundamental experiences of human existence and the emotional responses these experiences may inherently raise. This is one reason, as already noted, why religious narratives, with their capacity for articulating a holistic view of human personhood, can be so significant in contributing to the restoration age. A restoration age requires stories with plots and characters that speak to the richness, complexity, and depth of the human relationship to the natural world.

Fifth, we need stories about how people and natural processes can be interwoven in ways that promote both human well-being and ecological health. There remains a persistent dichotomy within environmental ethics and contemporary culture that suggests that society must choose between fostering human justice and well-being on the one hand and land's health and integrity on the other hand. Such binary thinking, however, obscures the basic idea that the whole problem of health, as Sir Albert Howard articulated in the early 1900s, intrinsically involves humans, animals, plants, and soil. The overarching goal of the restoration age must be health, in the fullest sense of the word, for people and land. To achieve this goal, we will need stories about particular communities, organizations, and institutions that are actively promoting models where human and natural systems are being intimately interwoven. The biophilic design and urban farming movements are good examples of the ways in which efforts toward human empowerment and well-being and ecological health and integrity can successfully coincide. The stories of characters enacting such models need to be told, and retold, as a way to counter and replace dominant narratives that pit human and ecological health against one another.

Sixth, stories for a restoration age need to include depictions of the sacredness of earth and the human experience of it. Encounters within the community of life on earth invite and necessitate a religious or spiritual response. Moreover, it is difficult to narrate our deepest experiences of the natural world without speaking in religious language. Artists, writers, dramatists, and musicians throughout history have depicted the human relationship to the natural world in ways that point to its spiritual elements. These depictions have often represented the religious idea that earth and its beings are characterized fundamentally by a unifying reality that illuminates the world with meaning and significance.

The sacred dimension of our connection with the natural world is a stream in the restorative story that has yet to flourish within scientific or religious narratives. This book has exhibited the difficulty involved with attempting to bridge or weave together religious, ethical, and scientific language about people and earth. Yet a more integrated language for describing the whole matter of people attempting to live more harmoniously with nature will not emerge and flourish until we speak and hear more of it—from scientists, conservationists, philosophers, theologians, religious leaders, and practitioners. As Thomas Berry and others have proposed, society may need scientist shamans or scientist poets who can narrate the evolutionary ecological story of the universe in a way that draws human beings into the depths of its mystery, complexity, ordering, and beauty. And it may also need prophetic ecologists who can articulate the wonder of the creation story and the healing of wounded land and spirit from within the context of particular religious communities.

I noted but one example of the way in which the new story of restoration might be narrated through a broader religious motif—that of redemption. Yet there are other religious narratives that may also be drawn upon and re-storied in ways that deepen and expand the human relationship to earth. Prophetic ecologists may narrate the ways in which modes of sacred presence have been wiped out where certain natural ecosystems and species have been wounded in a community's region; conversely, they may tell about places of healing where the community has been inspired to join in land's regeneration. Prophetic ecologists may bring the liturgy outdoors, making rituals and telling stories about the struggles and joys involved with attempting to live in communion with nature and others. They may draw out and accentuate the earthiness of the sacraments with their land-based elements. Liturgical calendars may tell the story of the seasons of the earth along with divine blessings for creation; sacred music and art can intimate nature's sounds and patterns. Metaphors of individual human salvation and transcendence can be intimately tied to those of earth's salvific pattern of perpetually regenerating life in and through death. The spiritual life and its pathways for achieving satisfaction and fulfillment in this life can be narrated with regard to the human relationship to a particular landed place.

Drawing out and animating religious narratives that speak specifically to earth's and humanity's healing will be one of Christianity's, and also other traditions', chief challenges in the restoration era. As in other historical periods, religious narratives may provide fuel for the development of stories that foster the self-fulfilling capacity of human and ecological life or they may contribute to the maintenance of stories that promote the self-destructive aspects of human beings in relation to earth and its beings. Neither function of religious narrative is inevitable; either path depends ultimately on the human capacity to choose a particular course of action. As global society enters the restoration age, religious traditions and communities will need to discern which narratives will best foster the health of humans together with the natural world.

CONCLUSION: RESTORING EARTH, RESTORED TO EARTH

The claim that people may become restored to earth in and through the process of restoring earth has been the fundamental thesis of this book. Further, I have argued that ecological restoration practice can serve as a paradigmatic context for developing a religious environmental ethic, as well as a broader environmental ethic. That is, ecological restoration may not only provide an experiential, vernacular activity for connecting people with the natural world, it may also provide a theoretical, methodological basis for better understanding the human relationship to nature and for generating a restorative environmental ethic.

Ecological restoration is distinctive as a social ecological practice because its meaning can be understood from various perspectives, including scientific and religious ones. Restoration may function primarily as practical ecological work, but it can also provide ritual, meaning-making activities. This dual potential of restoration is significant, especially when we consider the importance of relating scientific, ethical, and religious outlooks in bringing about cultural environmental change.

Social, ecological, religious, and ethical values and virtues are importantly shaped by concrete experiences and actions in relation to nature and others. These are experiences and actions that ecological restoration is well poised to provide. Restoration-based activities and the experiences they may reveal have multidimensional meanings for nature—including nature's integrity, historicity, interrelatedness, self-healing capacity, wildness, and sacredness.

Beyond this, acts of ecological restoration can yield personal and communal experiences of transformation and renewal in relation to damaged and healing land. They may generate ecological spiritual experiences characterized by a decentered and then recentered sense of selfhood, as well as by a communal sense of satisfaction and fulfillment in relation to particular natural lands and places. Moreover, ecological restoration practice both requires and serves as the basis for developing a spirituality of environmental action and an ethic of land partnership in which people and land become interwoven in active, cooperative, meaningful ways.

Ecological restoration and its practices can also shape good ecological community. Where the ethical principles of recognition, participation, and empowerment are realized, ecological restoration can provide an ideal model for building restorative communities of place. Restoration communities are places characterized by the ecological virtues of cohabitation, variety, wildness, sensuousness, publicity, and celebration. Although an ideal, restorative communities are rooted in the actions of groups of people who are already working to regenerate land-based social and ecological values in their particular contexts.

Restorative actions, along with the narratives that accompany them, will provide the fuel for the re-storying of society and the building of a restoration age. The re-storying of cultural narratives about nature and humanity will require both scientific and religious narratives of connection between people and earth. Stories about the regeneration of ecosystems, communities, and human beings will need to coalesce, for the restoration of earth's outer landscape is intimately linked with the restoration of humanity's inner landscape.

Narratives that become meaningful and take root within communities are not simply handed down from philosophers or theologians on high and then adopted uncritically by the moral community that receives them. Stories become formative for groups of people as they become their own, tested and shaped, illuminated and revised, according to the realities of day-to-day experience and life. Theorists may engage, give expression to, critique, and reformulate these storied interpretations of existence, of which their own lives are a part, as I have attempted to do here. Yet the interpretations need to be spoken back into the community, testing by trial and error their accuracy and adequacy, their power to interpret, express, and resonate with the experiences and norms that people already know.

Re-storying, as we have seen, is inherently a dynamic, evolving endeavor. There is, after all, changing land and biological life and a lack of wholeness in the human spirit. What I have leaned toward and argued throughout this book, nonetheless, is that wholeness or health in land and spirit is possible and that restorative actions can help sustain and foster this possibility. We can create narratives that are intimately attuned to the landscape within and without, to the healing ways of the personal heart and of the natural world. We can restore and re-story life in community in which we are more deeply connected to particular places and others. We can restore our own selves to earth, physically and spiritually, in and through the process of restoring earth. The restoration age is being fostered by thousands of communities of people engaged in acts of healing. It is through their stories—of their actions and their experiences of connection—that a broader movement of restoration, of land and spirit, will grow.

NOTES

1. Thomas Berry, *Dream of the Earth*, 42.
2. Ibid.
3. See, for example, Freyfogle, *Agrarianism and the Good Society*; Kellert, *Building for Life*; and Peterson, *Being Human*.
4. See, for example, environmental psychologist Louis Chawla's book *In the First Country of Places*. See also Dunlap and Kellert, *Companions in Wonder*.
5. On this idea see, for example, MacIntyre, *After Virtue*, 211–21.
6. Peterson, *Being Human*, 18.
7. Merchant, *Reinventing Eden*, 202.
8. Thayer, *LifePlace*, 104.
9. On this idea, see the collection of essays in Low and Altman, *Place Attachment*.
10. Merchant, *Reinventing Eden*, 202.
11. Ibid., 203.
12. Most basically, see again Lynn White's indictment of Christianity as the historical root of the current environmental crisis: "Historical Roots." Additionally, see the debate on the role of religion in constructing an environmental ethic between J. Baird Callicott and Bron Taylor in Taylor, "On Sacred or Secular Ground?"
13. The additional three problems Gottlieb notes are that "religion, in essence, is undemocratic and oppressive," that "involvement in politics is bad for religion," and that "religion has become increasingly irrelevant to modern life, so a religious environmentalism is not needed and will make no real contribution." See his *A Greener Faith*, 57–80.
14. Ibid., 59.
15. See Spencer's essay "Restoring Earth, Restored to Earth," 428.
16. Lynn White made this argument early on in his "The Historical Roots."
17. On these charges, see, for example, Thomas Berry, *Dream of the Earth*, 109–22.
18. On this critique, see Northcott, "Scottish Highlands," 382–99.
19. On redemption's role as a central metaphor within Western culture, see Higgs's reference to a cultural notion of redemption in his "What Is Good Ecological Restoration?" 342.
20. See, for example, George Kehm's theological treatment of it in his piece, "The New Story," in Hessel, *After Nature's Revolt*, 89–108.

21. It must be noted that these are major, though not universal ways, of construing redemption in the Christian tradition.
22. See Rasmussen's *Earth Community*, 256.
23. Willis Jenkins uses this phrase in an example of a sermon that Reformed church ministers might give to preach about the degradation and restoration of the Muskegon River in central Michigan. See Jenkins, *Ecologies of Grace*, 233–34.
24. Some Hebrew Bible scholars such as Ellen Davis question whether the Sabbatical laws were actually practiced in ancient Israel. Nonetheless, that they are part of the Hebrews' faith story reflects their centrality in the overall ethic of the community. Further, that Jesus reportedly references them in the Christian scriptures (e.g., "The year of the Lord's favor," Luke 4:19) suggests their historical and spiritual significance as well.
25. Ivone Gebara comments on this interconnection: "In an act of aggression against nature, the real target is the local human inhabitants. So there is no direct intention to destroy the fauna, flora, or fish in the rivers. Despite this, however, aggression against human beings also becomes aggression against all of nature, and the latter aggression is used as a weapon against human beings." *Longing for Running Water*, 27.
26. An agenda for a restoration age advances and goes beyond Thomas Berry's agenda for an ecological age, as stated in this chapter's introduction. See his *Dream of the Earth*, 65–69.
27. On this idea, see Eric Higgs's treatment of Leo Marx's term "mode" applied to ecological restoration in *Nature by Design*, 216.

BIBLIOGRAPHY

∾

Angermeier, P. L, and J. R. Karr. "Biological Integrity versus Biological Diversity as Policy Directives." *Bioscience* 44: 690–97.

Aquinas, Thomas. *Summa Theologica* I-II. 91–94. See Thomas Aquinas.

Aristotle. *The Nicomachean Ethics*. Translated by David Ross. Oxford and New York: Oxford University Press, 1925, 1980.

Atler, Joseph S. *Yoga in Modern India: The Body between Science and Philosophy*. Princeton: Princeton University Press, 2004.

Audi, Robert, ed. *Cambridge Dictionary of Philosophy*, 2nd ed. Cambridge: Cambridge University Press, 1999.

Augustine. *On Free Choice of the Will*. Translated by Anna S. Benjamin and L. H. Hackstaff. Upper Saddle River, NJ: Prentice Hall, 1946.

Bass, Dorothy C., ed. *Practicing Our Faith: A Way of Life for a Searching People*. San Francisco: Jossey-Bass Publishers, 1997.

Beatley, Timothy. *Native to Nowhere: Sustaining Home and Community in a Global Age*. Washington, DC: Island Press, 2004.

Bell, Catherine. *Ritual Theory, Ritual Practice*. Oxford: Oxford University Press, 1992.

Bellah, Robert, Richard Mardsen, William M. Sullivan, Ann Swidler, and Steve Tipton. *Habits of the Heart: Individualism and Commitment in American Life*. Berkeley and Los Angeles: University of California Press, 1985.

Benjamin, Jessica. *The Bonds of Love: Psychoanalysis, Feminism and the Problem of Domination*. London: Virago, 1988.

Berg, Peter, ed. *Reinhabiting a Separate Country: A Bioregional Anthology of Northern California*. San Francisco: Planet Drum Books, 1978.

Berg, Peter, and Raymond Dasmann. "Reinhabiting California," in Berg, *Reinhabiting a Separate Country*.

Berry, Robert James, ed. *Environmental Stewardship: Critical Perspectives, Past and Present*. London, New York: T&T Clark, 2006.

Berry, Thomas. *The Dream of the Earth*. San Francisco: Sierra Club Books, 1988.

Berry, Wendell. *Another Turn of the Crank*. Washington, DC: Counterpoint, 1995.

———. *The Unsettling of America*. San Francisco: Sierra Club Books, 1977.

Borgmann, Albert. *Technology and the Character of Contemporary Life: A Philosophical Inquiry*. Chicago: University of Chicago Press, 1984.

Bouma-Prediger, Steven. *For the Beauty of the Earth: A Christian Vision for Creation Care*. Grand Rapids, MI: Baker Academic Press, 2001.

Bouma-Prediger, Steven, and Peter Bakken, eds. *Evocations of Grace: Writings on Ecology, Theology, and Ethics*. Grand Rapids, MI: William B. Eerdmans Publishing Company, 2000.

Boyce, James K., Sunita Narain, and Elizabeth A. Stanton, eds. *Reclaiming Nature: Environmental Justice and Ecological Restoration*. London, New York, and Delhi: Anthem Press, 2007.

Briggs, Beatrice. "Help Wanted: Scientists-Shamans and Eco-Rituals." *Restoration and Management Notes* 12, no. 1 (1994): 24.

Brown, Raphael, trans. *The Little Flowers of Saint Francis*. New York: Doubleday Books, 1958.

Bullard, Robert D., ed. *The Quest for Environmental Justice: Human Rights and the Politics of Pollution*. San Francisco: Sierra Club Books, 2005.

Cabin, Robert J. *Intelligent Tinkering: Bridging the Gap between Science and Practice*. Washington, DC: Island Press, 2011.

Callicott, J. Baird. "Holistic Environmental Ethics and the Problem of Ecofascism." In *Environmental Philosophy: From Animal Rights to Radical Ecology*, edited by J. Baird Callicott, Michael E. Zimmerman, George Sessions, Karen J. Warren, and John Clark. Upper Saddle River, NJ: Prentice Hall, 2001.

———. "La Nature est morte, vive la nature!" *Hastings Center Report* 22, no. 5 (1992): 16–23.

Calvin, John. *Institutes of the Christian Religion*. Translated by Ford Lewis Battles and edited by John T. McNeil. Philadelphia: Westminster Press, 1960.

Chawla, Louis. *In the First Country of Places: Nature, Poetry, and Childhood Memory*. Albany: State University of New York Press, 1994.

Davis, Mark, and Lawrence Slobodkin. "The Science and Values of Restoration Ecology," *Restoration Ecology* 12, no. 1 (2004): 1–3.

Day, Dorothy. *Loaves and Fishes*. Maryknoll, NY: Orbis Books, 1963.

de Michelis, Elizabeth. *A History of Modern Yoga: Patanjali and Western Esotericism*. London and New York: Continuum, 2004.

Duncan, Dayton. "The National Parks: America's Best Idea," Film series. Directed by Ken Burns and Dayton Duncan (Public Broadcasting Service, 2009), available at http://www.pbs.org/nationalparks/.

Dunlap, Julie, and Stephen R. Kellert, eds. *Companions in Wonder: Children and Adults Exploring Nature Together*. Cambridge, MA: MIT Press, 2011.

Elliot, Robert. "Faking Nature." *Inquiry* 25 (1982).

Emerson, Ralph Waldo. "Nature" (1844). In *Emerson's Essays*, with an introduction by Irwin Edman, 380–401. New York: Thomas Crowell, 1926, 1961.

Falk, Donald A., Margaret A. Palmer, and Joy B. Zedler, eds. *Foundations of Restoration Ecology*. Washington, DC: Island Press, 2006.

Farley, Margaret. *Just Love: A Framework for Christian Sexual Ethics*. New York and London: Continuum, 2006.

———. "Religious Meanings for Nature and Humanity." In *The Good in Nature and Humanity: Connecting Science, Religion, and Spirituality with the Natural World*, edited by Stephen R. Kellert and Timothy J. Farnham. Washington, DC: Island Press, 2001.

Farnham, Timothy J., and Stephen R. Kellert, "Building the Bridge: Connecting Science, Religion, and Spirituality with the Natural World." In *The Good in Nature and Humanity: Connecting Science, Religion, and Spirituality with the Natural World*, edited by Stephen R. Kellert and Timothy J. Farnham. Washington, DC: Island Press, 2002.

Flader, Susan L., and J. Baird Callicott, eds. *River of the Mother of God and other Essays by Aldo Leopold*. Madison: University of Wisconsin Press, 1991.

Freyfogle, Eric T. *Agrarianism and the Good Society: Land, Culture, Conflict, and Hope*. Lexington: The University Press of Kentucky, 2007.

Fromm, Erich. *Escape from Freedom*. New York: Henry Holt and Company, 1941, 1969.

Gebara, Ivone. *Longing for Running Water: Ecofeminism and Liberation*. Minneapolis, MN: Augsburg Press, 1999.

Glacken, Clarence. *Traces on the Rhodian Shore: Nature and Culture in Western Thought from Ancient Times to the End of the Eighteenth Century*. Berkeley: University of California Press, 1967.

Gottlieb, Roger S. *A Greener Faith: Religious Environmentalism and Our Planet's Future*. London and Oxford: Oxford University Press, 2006.

———. *A Spirituality of Resistance: Finding a Peaceful Heart and Protecting the Earth*. New York: The Crossroad Publishing Company, 1999.

———. "The Transcendence of Justice and the Justice of Transcendence: Mysticism, Deep Ecology, and Political Life." *Journal of the American Academy of Religion* 67, no. 1 (1999): 149–66.

Gould, Rebecca Kneale. *At Home in Nature: Modern Homesteading and Spiritual Practice in America*. Berkeley: University of California Press, 2005.

Gustafson, James. *Ethics from a Theocentric Perspective*. Vol. 1. Chicago: University of Chicago Press, 1981.

———. *A Sense of the Divine: The Natural Environment from a Theocentric Perspective*. Cleveland: Pilgrim Press, 1994.

Hall, John Douglas. *Professing the Faith: Christian Theology in a North American Context*. Minneapolis, MN: Augsburg Fortress Press, 1993.

Hall, Marcus. "American Nature, Italian Culture: Restoring the Land in Two Continents." PhD diss., University of Wisconsin–Madison, 1999.

———. "Co-Workers with Nature: The Deeper Roots of Restoration." *Restoration and Management Notes* 15, no. 2 (1997): 173–78.

Haraway, Donna. "The Actors Are Cyborg, Nature Is Coyote, and the Geography Is Elsewhere: Postscript to Cyborgs at Large." In *Technoculture*, edited by Constance Penley and Andrew Ross. Minneapolis: University of Minnesota Press, 1991.

———. *Simians, Cyborgs, and Women: The Reinvention of Nature.* New York and London: Routledge, 1991.

Harrison, Beverly Wildung. *Justice in the Making: Feminist Social Ethics,* edited by Elizabeth M. Bounds, Pamela K. Brubaker, Jane E. Hicks, Marilyn J. Legge, Rebecca Todd Peters, and Traci C. West. Louisville and London: Westminster John Knox Press, 2004.

Hauerwas, Stanley. *Character in Christian Life: A Study in Theological Ethics.* South Bend, IN: University of Notre Dame Press, 1989.

Heerwagen, Judith H., and Gordon H. Orians. "The Ecological World of Children." In *Children and Nature: Psychological, Sociocultural, and Evolutionary Investigations,* edited by Stephen R. Kellert and Peter H. Kahn Jr. Cambridge, MA: MIT Press, 2002.

Hessel, Dieter T., ed. *After Nature's Revolt: Eco-Justice and Theology.* Minneapolis, MN: Augsburg Press, 1992.

Hessel, Dieter T., and Rosemary Radford Ruether, eds. *Christianity and Ecology: Seeking the Well-Being of Earth and Humans.* Cambridge, MA: Harvard University Press, 2000.

Higgs, Eric. *Nature by Design: People, Natural Process, and Ecological Restoration.* Cambridge, MA: MIT Press, 2003.

———. "What Is Good Ecological Restoration?" *Conservation Biology* 11, no. 2 (April 1997).

Holland, Karen M. "Restoration Rituals: Transforming Workday Tasks into Inspirational Rites." *Restoration and Management Notes* 12, no. 2 (1994): 121–25.

House, Freeman. *Totem Salmon: Life Lessons from another Species.* Boston: Beacon Press, 1999.

Jenkins, Willis. *Ecologies of Grace: Environmental Ethics and Christian Theology.* New York and Oxford: Oxford University Press, 2007.

———. "After Lynn White: Religious Ethics and Environmental Problems." *Journal of Religious Ethics* 37, no. 2 (2009).

Jones, Serene. *Feminist Theory and Theology: Cartographies of Grace.* Minneapolis, MN: Fortress Press, 2000.

Jordan, William R., III. "Ghosts in the Forest," *Restoration and Management Notes* 11, no. 1 (1993): 3–4.

———. *The Sunflower Forest: Ecological Restoration and the New Communion with Nature.* Berkeley: University of California Press, 2003.

Jordan, William R., III, and George M. Lubick. *Making Nature Whole: A History of Ecological Restoration.* Washington, DC: Island Press, 2011.

Jordan, William, III, Nathaniel F. Barrett, Kip Curtis, Liam Henneghan, Randall Honold, Todd LeVasseur, Anna Peterson, Leslie Paul Thiel, and Gretel Van Wieren. "Foundations of Conduct: A Theory of Values and Its Implications for Environmentalism." *Environmental Ethics* 34 (2012): 291–312.

Journal of the American Academy of Religion. Special issue on "Aquatic Nature Religion." Vol. 75, no. 4 (December 2007).

Katz, Eric. "The Big Lie: Human Restoration of Nature." *Research in Philosophy and Technology* 12 (1992).

Kearns, Laurel, and Catherine Keller, eds. *Ecospirit: Religions and Philosophies for the Earth.* New York: Fordham University Press, 2007.

Kehm, George. "The New Story: Redemption as Fulfillment of Creation." In *After Nature's Revolt: Eco-Justice and Theology*, edited by Dieter Hessel. Minneapolis, MN: Augsburg Fortress Press, 1992.

Keller, Catherine. *Face of the Deep: A Theology of Becoming.* London and New York: Routledge, 2003.

Keller, Evelyn Fox. *Reflections on Gender and Science.* New Haven, CT: Yale University Press, 1985.

Kellert, Stephen R. *Kinship to Mastery: Biophilia in Human Evolution and Development.* Washington, DC: Island Press, 1997.

———. *Building for Life: Designing and Understanding the Human-Nature Connection.* Washington, DC: Island Press, 2005.

———. *Birthright: People and Nature in the Modern World.* New Haven, CT: Yale University Press, 2012.

Kellert, Stephen R., and Timothy J. Farnham, eds. *The Good in Nature and Humanity: Connecting Science, Religion, and Spirituality with the Natural World.* Washington, DC: Island Press, 2002.

Kellert, Stephen R., and Peter H. Kahn Jr., eds. *Children and Nature: Psychological, Sociocultural, and Evolutionary Investigations.* Cambridge, MA: MIT Press, 2002.

Kellert, Stephen R., and James Gustave Speth, eds. *The Coming Transformation: Values to Sustain Human and Natural Communities.* New Haven, CT: Yale School of Forestry and Environmental Studies, 2009.

Kellert, Stephen R., and Edward O. Wilson, eds. *The Biophilia Hypothesis.* Washington, DC: Island Press, 1998.

Lane, Belden C. *Ravished by Beauty: The Surprising Legacy of Reformed Spirituality.* New York: Oxford University Press, 2011.

Lang, Julian. Introduction, *To the American Indian: Reminiscences of a Yurok*, by Lucy Thompson. San Leandro, CA: Heyday Books, 1991.

Leiserowitz, Anthony A., and Lisa Fernandez. "Toward a New Consciousness: Values to Sustain Human and Natural Communities." In *A Synthesis of Insights and Recommendations from the 2007 Yale F&ES Conference*, 2007.

Leopold, Aldo. *A Sand County Almanac: And Sketches Here and There.* New York and Oxford: Oxford University Press, 1949, 1987.

———. "The Farmer as a Conservationist." *American Forests* 45 (1939): 294–99.

Light, Andrew. "Ecological Citizenship: The Democratic Promise of Restoration." In *The Humane Metropolis: People and Nature in the Twenty-first Century City*, edited by R. Platt, 176–89. Amherst: University of Massachusetts Press, 2005.

———. "Ecological Restoration and the Culture of Nature: A Pragmatic Perspective." In *Restoring Nature*, edited by Paul H. Gobster and R. Bruce Hull, 49–70. Washington, DC: Island Press.

———. "Restoration, the Value of Participation, and the Risks of Professionalization." In *Restoring Nature: Perspectives from the Social Sciences and Humanities*, edited by Paul H. Gobster and Bruce Hull. Washington DC: Island Press, 2000.

Light, Andrew, and Eric Higgs. "The Politics of Ecological Restoration." *Environmental Ethics* 18 (1996): 227–47.

Light, Andrew, and Holmes Rolston III, eds. *Environmental Ethics: An Anthology*. Malden, MA: Blackwell, 2003.

Low, Setha, and Irwin Altman, "Place Attachment: A Conceptual Inquiry." In *Place Attachment*, edited by Irwin Altman and Setha M. Low. New York: Plenum Press, 1992.

———, eds. *Place Attachment*. New York: Plenum Press, 1992.

Maathai, Wangari. "A Billion Trees, A Singular Voice." *Reflections* (Spring 2007): 5.

———. *The Green Belt Movement: Sharing the Approach and the Experience*. New York: Lantern Books, 2004.

MacIntyre, Alasdair. *After Virtue: A Study in Moral Theory*. Notre Dame: University of Notre Dame Press, 1981.

Marsh, George Perkins. *Man and Nature*. Cambridge, MA: Harvard University Press, 1864, 1965.

Martinez, Dennis Rogers. "Northwestern Coastal Forests: The Sinkyone Intertribal Park Project." *Restoration and Management Notes* 10, no. 1 (Summer 1992): 64–69.

McFague, Sallie. *Super Natural Christians: How We Should Love Nature*. Minneapolis, MN: Augsburg Fortress Press, 1997.

———. *Life Abundant: Rethinking Theology and Economy for a Planet in Peril*. Minneapolis, MN: Augsburg Fortress Press, 2001.

McKibben, Bill. *The End of Nature*. New York: Anchor Books, 1990.

McQuillan, Alan. "Defending the Ethics of Ecological Restoration." *Journal of Forestry* 96, no. 1 (1998): 27–31.

Meekison, Lisa, and Eric Higgs. "Rites of Spring (and Other Seasons): The Ritualization of Restoration." *Restoration and Management Notes* 16, no. 1 (1998).

Meisel, Anthony C., and M. L. del Mastro, trans. *Benedict, The Rule of Saint Benedict*. New York: Doubleday, 1975.

Mendelson, Jon. Stephen P. Aultz, and Judith Dolan Mendelson. "Carving Up the Woods: Savanna Restoration in Northeastern Illinois." In Throop, *Environmental Restoration*.

Merchant, Carolyn. *Reinventing Eden: The Fate of Nature in Western Culture*. New York: Routledge, 2004.

Miles, Irene, William C. Sullivan, and Frances E. Kuo. "Psychological Benefits of Volunteering for Restoration Projects." *Ecological Restoration* 18, no. 4 (2000): 218–27.

Mill, John Stuart. "Nature." In *Essays on Ethics, Religion and Society*, from *Collected Works of John Stuart Mill*, vol. 10, 372–402. Toronto: University of Toronto Press, 1969.

Mills, Stephanie. *In Service of the Wild: Restoring and Reinhabiting Damaged Land*. Boston: Beacon Press, 1995.

Moltmann, Jurgen, *God in Creation: A New Theology of Creation and the Spirit of God.* San Francisco: Harper and Row, 1985.

Norris, Kathleen. *Dakota: A Spiritual Autobiography.* New York: Houghton-Mifflin, 1993.

Northcott, Michael S. *The Environment and Christian Ethics.* Cambridge: Cambridge University Press, 1996.

———. "From Environmental U-topianism to Parochial Ecology: Communities of Place and the Politics of Sustainability," *Ecotheology* 8 (2000): 71–85.

———. "Wilderness, Religion, and Ecological Restoration in the Scottish Highlands," *Ecotheology* 10 (2005): 382–99.

O' Brien, Kevin J. *An Ethics of Biodiversity: Christianity, Ecology, and the Variety of Life.* Washington, DC: Georgetown University Press, 2010.

O'Neill, J. "Time, Narrative and Environmental Politics." In *Ecological Community*, edited by Roger Gottlieb. London: Routledge 1997.

O'Neill, J., and A. Holland. "Two Approaches to Biodiversity Value." In *Cultural and Spiritual Values in Biodiversity*, edited by D. Posey. London: UNEP, 1999.

Palmer, Margaret A., Donald A. Falk, and Joy B. Zedler. "Ecological Theory and Restoration Ecology." In *Foundations of Restoration Ecology*, edited by Donald A. Falk, Margaret A. Palmer, and Joy B. Zedler. Washington, DC: Island Press, 2006.

Peters, Rebecca Todd. *In Search of the Good Life: The Ethics of Globalization.* New York and London: Continuum, 2004.

Petersen, David. "Hunting for Spirituality." In *The Good in Nature and Humanity*, edited by Stephen R. Kellert and Timothy J. Farnham. Washington, DC: Island Press, 2001.

Peterson, Anna. *Being Human: Ethics, Environment, and Our Place in the World.* Berkeley: University of California Press, 2001.

———. "Talking the Walk: A Practice-Based Environmental Ethic as Grounds for Hope." In *Ecospirit: Religions and Philosophies for the Earth*, edited by Laurel Kearns and Catherine Keller, 45–62. New York: Fordham University Press, 2007.

Plumwood, Val. *Environmental Culture: The Ecological Crisis of Reason.* London: Routledge, 2002.

Rasmussen, Larry. *Earth Community, Earth Ethics.* Maryknoll, NY: Orbis, 1996.

———. "Returning to Our Senses: The Theology of the Cross as a Theology for Eco-Justice." In *After Nature's Revolt: Eco-Justice and Theology*, edited by Dieter T. Hessel. Minneapolis, MN: Augsburg Press, 1992.

Ray, Janisse. *Pinhook: Finding Wholeness in a Fragmented Land.* White River Junction, VT: Chelsea Green Publishing, 2005.

Reece, Erik. "Reclaiming a Toxic Legacy through Art and Science: Putting Art to Work." *Orion*, November–December 2007.

Regan, Tom. "Animal Rights." In Light and Rolston, *Environmental Ethics.*

Rolston, Holmes, III. "Restoration." In *Environmental Restoration: Ethics, Theory, and Practice*, edited by William Throop. Amherst, NY: Humanity Books, 2000.

———. *Environmental Ethics: Duties to and Values in the Natural World.* Philadelphia: Temple University Press, 1988.

———. *Genes, Genesis, and God: Values and Their Origins in Nature and Human History.* Cambridge: Cambridge University Press, 1999.

———. "Naturalizing and Systematizing Evil." In *Is Nature Ever Evil? Religion, Science, and Value,* edited by Willem B. Drees, 67–89. London: Routledge, 2003.

———. "Value in Nature and the Nature of Value." In Light and Rolston, *Environmental Ethics.*

Roof, Wade Clark. *A Generation of Seekers.* San Francisco: Harper, 1993, 75–76.

Ruether, Rosemary Radford. "The Biblical Vision of the Ecological Crisis." *Christian Century,* November 22, 1978, 1132.

———. *Gaia and God: An Ecofeminist Theology of Earth Healing.* San Francisco: Harper Collins, 1992.

———, ed. *Women Healing Earth: Third World Women on Ecology, Feminism, and Religion.* Maryknoll, NY: Orbis Books, 1996.

Sagoff, Mark. "Do Non-Native Species Threaten the Natural Environment?" *Journal of Agricultural and Environmental Ethics* 18 (2005): 215–36.

Santmire, H. Paul. "Partnership with Nature according to the Scriptures: Beyond the Theology of Stewardship." *Christian Scholar's Review* 32:4 (Summer 2003): 381–412.

———. *Ritualizing Nature: Renewing Christian Liturgy in a Time of Crisis.* Minneapolis, MN: Fortress Press, 2008.

Schlosberg, David. *Environmental Justice and the New Pluralism.* Oxford: Oxford University Press, 1999.

Science and Policy Working Group of the Society for Ecological Restoration, "The SER International Primer on Ecological Restoration," available at http://www.ser.org/content/ecological_restoration_primer.asp, 3.

Shore, Debra. "Controversy Erupts over Restoration in Chicago Area." *Restoration and Management Notes* 15, no. 1 (1997).

Shrader-Frechette, Kristin. *Environmental Justice: Creating Equality, Reclaiming Democracy.* New York: Oxford University Press, 2002.

Sideris, Lisa H. *Environmental Ethics, Ecological Theology, and Natural Selection.* New York: Columbia University Press, 2003.

Simberloff, Daniel. "Non-Native Species Do Threaten the Natural Environment!" *Journal of Agricultural and Environmental Ethics* 18 (2005): 595–607.

Singer, Peter. "Not for Humans Only." In Light and Rolston, *Environmental Ethics.*

Sittler, Joseph. "Called to Unity." In *Evocations of Grace: Writings on Ecology, Theology, and Ethics,* edited by Steven Bouma-Prediger and Peter Bakken. Grand Rapids, MI: William B. Eerdmans Publishing Company, 2000.

Smith, Jonathan Z. *To Take Place: Towards Theory in Ritual.* Chicago Studies in the History of Judaism, edited by Jacob Neusner et al. Chicago: University of Chicago Press, 1987, 1992.

Snyder, Samuel. "New Streams of Religion: Fly Fishing as a Lived, Religion of Nature." *Journal of the American Academy of Religion* 75, no. 4 (December 2007): 896–922.

Soule, Michael E., and Gary Lease, eds. *Reinventing Nature?: Responses to Postmodern Deconstruction.* Washington, DC: Island Press, 1995.

Spencer, Daniel T. "Restoring Earth, Restored to Earth: Toward an Ethic for Reinhabiting Place." In *Ecospirit: Religions and Philosophies for the Earth*, edited by Laurel Kearns and Catherine Keller, 415–32. New York: Fordham University Press, 2007.

Stange, Mary Zeiss, *Woman the Hunter*. Boston: Beacon Press, 1997.

Stegner, Wallace. *The American West as Living Space*. Ann Arbor: University of Michigan Press, 1987.

Stockinger, Jacob. "Classical Music Review: Saving the Land and Hearing Great Music Make for an Outstanding Double-Header at the Prairie Rhapsody Benefit Concert." *The Well-Tempered Ear* (blog). July 19, 2011. http://welltempered .wordpress.com/2011/07/19/classical-music-review-music-preservation-and-land-conservation-make-an-outstanding-match-at-the-prairie-rhapsody-concert/.

Strauss, Sarah. *Positioning Yoga: Balancing Acts across Cultures*. Oxford and New York: Berg, 2005.

Swimme, Brian Thomas, and Mary Evelyn Tucker. *Journey of the Universe*. New Haven, CT: Yale University Press, 2011.

Sylvan (Routley), Richard. "Is There a Need for a New, an Environmental Ethic?" In Light and Rolston, *Environmental Ethics*.

Taylor, Bron. *Dark Green Religion: Nature Spirituality and the Planetary Future*. Berkeley: University of California Press, 2010.

———, ed. *Encyclopedia of Religion and Nature*. New York: Continuum, 2005.

———. "Deep Ecology and Its Social Philosophy: A Critique." In *Beneath the Surface: Critical Essays on Deep Ecology*, edited by Eric Katz, Andrew Light, and David Rothenberg, 269–99. Cambridge, MA: MIT Press, 2000.

———. Introduction to *Encyclopedia of Religion and Nature*.

———. "On Sacred or Secular Ground? Callicott and Environmental Ethics," *Worldviews: Environment, Culture, Religion* 1 (1997): 99–111.

———. "Surfing into Spirituality and a New, Aquatic Nature Religion." *Journal of the American Academy of Religion* 75, no. 4 (December 2007): 923–51.

———. "Focus Introduction: Aquatic Nature Religion," *Journal of the American Academy of Religion* 75, no. 4 (December 2007): 867.

Taylor, Charles. "The Politics of Recognition." In *Multiculturalism*, edited by Amy Gutmann. Princeton, NJ: Princeton University Press, 1992.

Taylor, Paul W. "The Ethics of Respect for Nature." In Light and Rolston, *Environmental Ethics*.

Taylor, Sarah McFarland. *Green Sisters: A Spiritual Ecology*. Cambridge and London: Harvard University Press, 2007.

Thayer, Robert L., Jr. *LifePlace: Bioregional Thought and Practice*. Berkeley: University of California Press, 2003.

Thomas Aquinas. *Introduction to St. Thomas Aquinas: The Summa Theologica, the Summa Contra Gentiles*. Edited by Anton C. Pegis. New York: Modern Library, 1948.

Throop, William. "Eradicating the Aliens: Restoration and Exotic Species." In *Environmental Restoration: Ethics, Theory, and Practice*, edited by William Throop. Amherst, NY: Humanity Books, 2000.

———, ed. *Environmental Restoration: Ethics, Theory, and Practice.* Amherst, NY: Humanity Books, 2000.

Tucker, Mary Evelyn. *Worldly Wonder: Religions Enter Their Ecological Phase.* Peru, IL: Carus Publishing Company, 2003.

Tucker, Mary Evelyn, and John Grim. "Series Forward." In *Christianity and Ecology*, edited by Hessel and Ruether, xviii.

———. "Introduction: The Emerging Alliance of the World Religions and Ecology." *Daedalus* (2001). Available at http://www.amacad.org/publications/fall2001/tucker-grim.aspx.

Turner, Frederick. "Bloody Columbus: Restoration and the Transvaluation of Shame into Beauty." *Restoration and Management Notes* 10, no. 1 (Summer 1992): 70–74.

Van Wieren, Gretel. "Children in the River." In *Companions in Wonder: Children and Adults Exploring Nature Together,* edited by Julie Dunlap and Stephen R. Kellert. Cambridge, MA: MIT Press, 2011.

———. "Ecological Restoration as Public Spiritual Practice," *Worldviews* 12 (2008): 237–54.

Wallace, Mark I. *Finding God in the Singing River: Christianity, Spirit, Nature.* Minneapolis, MN: Augsburg Fortress Press, 2005.

———. "Sacred-Land Theology: Green Spirit, Deconstruction, and the Question of Idolatry in Contemporary Earthen Christianity." In *Ecospirit: Religions and Philosophies for the Earth*, edited by Laurel Kearns and Catherine Keller, 291–314. New York: Fordham University Press, 2007.

Walzer, Michael. *Spheres of Justice: A Defense of Pluralism and Equality.* New York: Basic Books, 1983.

White, Lynn, Jr. "The Historical Roots of Our Ecological Crisis." *Science* 155 (1967): 1203–7.

Winterhalder, Keith, Andre F. Clewell, and James Aronson. "Values and Science in Ecological Restoration—Response to Davis and Slobodkin." *Restoration Ecology* 12, no. 1 (2004): 4.

Worster, Donald. *The Wealth of Nature: Environmental History and the Ecological Imagination.* New York: Oxford Unviersity Press, 1993.

Wuthnow, Robert. *After Heaven: Spirituality in America since the 1950s.* Berkeley: University of California Press, 1998.

Young, Iris Marion. *Justice and the Politics of Difference.* Princeton, NJ: Princeton University Press, 1990.

Zimmerman, Michael E., J. Baird Callicott, and Karen J. Warren, eds. *Environmental Philosophy: From Animal Rights to Radical Ecology.* Saddleback, NJ: Prentice Hall, 2004.

INDEX

Holy Wisdom Monastery, 134, 147, 152–55, 156, 164–65, 180

Hooker, Richard, 112n76

House, Freeman, 95, 102, 157; and bioregionalism, 44, 46; and Mattole Valley project, 157–58, 159–60; on reinhabitation, 15, 21, 45

Howard, Sir Albert, 26, 183

human beings, 25, 41, 172; human-to-human relationships, 106; need for redemption by, 176; needs and desires, 183; and nonhuman beings, 20–21, 120–22, 130

human-nature relationship, 59–60, 149; anthropocentrism in, 8, 29–30n24, 59, 161; Bible on, 178, 188n25; Christian spiritual view on, 101; ecological restoration and, 2, 3, 115, 128, 130, 171, 180; holistic view of, 176; human dependency on nature, 95; human domination in, 48, 56, 59, 60, 159; humans as part of nature, 42–43; land and, 29, 65, 107, 170, 172, 173; nonhuman natural beings and, 20–21, 120–22, 130; as partnership, 107, 134, 173; re-storying of, 171–74, 180, 182–84. *See also* nature

hunting, 90, 131

Hurricane Katrina, 135

Illick, Marty, 84, 85, 156

imagination, 172

indigenous peoples: biocultural restoration and, 43, 44; ecological restoration by, 37, 39, 41, 52n5, 70, 93; and environmental decision making, 126; and redistributive justice, 125

Indigenous Peoples' Restoration Network (IPRN), 15, 23, 43, 52n4, 53n44

indigenous species, 120, 121, 122

integrity, ecological, 40, 67–69, 82n53, 147

interrelatedness, 74–75

invasive species, 122, 140–41n26

Ishmael, 183

Jenkins, Willis, 22, 188n23

Jensen, Jens, 37

Jesus Christ, 150, 175

Job, 183

Joliet, IL, 2, 147

Jones, Serene, 82n42

Jordan, William, 14, 31n59, 38, 52n5, 66, 131, 148; on community, 122, 149; on nature, 3, 78, 148, 161; and *Restoration and Management Notes*, 3, 38; as restoration visionary, 16; on ritual performance and tradition, 22, 46, 47, 102, 148, 149–50, 151; *The Sunflower Forest*, 16

justice, 140n18, 183; ecological restoration and, 14, 23–24, 123–24, 125, 126; and environmental movement and, 117–18; and nature, 23, 117; and participation, 126, 133

Kant, Immanuel, 62

Karr, James, 68

Katz, Eric, 10, 12, 48, 56, 58–60, 62, 80n10, 115

Kay, James, 40

Keller, Catherine, 101

Keller, Evelyn Fox, 75